EVERYONE IS AFRICAN

EVERYONE IS AFRICAN
HOW SCIENCE EXPLODES THE MYTH OF RACE

DANIEL J. FAIRBANKS

Prometheus Books

59 John Glenn Drive
Amherst, New York 14228

Published 2015 by Prometheus Books

Cover image © Bigstock
Cover design by Grace M. Conti-Zilsberger

Unless otherwise noted, all images are original drawings by the author.

All DNA sequences are derived from public databases of the
National Center for Biotechnology Information and are in the public domain.

Inquiries should be addressed to
Prometheus Books
59 John Glenn Drive
Amherst, New York 14228
VOICE: 716–691–0133
FAX: 716–691–0137
WWW.PROMETHEUSBOOKS.COM

19 18 17 16 15   5 4 3 2 1

Library of Congress Cataloging-in-Publication Data Pending

ISBN 978-1-63388-018-4 (paperback) ISBN 978-1-63388-019-1 (ebook)

Printed in the United States of America

CONTENTS

PREFACE

Few subjects elicit such powerful emotion as race and racism. The history of racism is filled with such horrific cruelty and abuse that much of it has been excluded from historical accounts. In recent years, however, several well-written and thoroughly researched works have candidly recounted the history of racism, and they document some of the most horrendous persecutions ever imposed upon large groups of people over multiple generations. As these more recent histories were being written, human geneticists were collecting evidence supporting one of the most significant discoveries in the history of science—evidence that the notion of distinct biological races is flawed.

It is not uncommon to hear or read statements like the following: "Race is a socially constructed concept, not a biological one;"[1] "In scientific terms, racial differences have no material significance;"[2] "Today, science is eroding the biological basis of the idea of race;[3] or "Race is an exceedingly slippery concept . . . while it is a biological fiction, it is nonetheless a social fact."[4]

Such statements seem to contradict the obvious: that children inherit the features we typically associate with race—especially skin, hair, and eye color—from their parents. Because that inheritance must be genetic, how can we legitimately say that race is social rather than biological? And how, then, can science (as the title of this book proclaims) explode the myth of race?

Statements like those quoted above usually appear as declarations of fact, with little or no contextual evidence or explanation to justify them. Even so, they are neither hollow nor dogmatic. Rather, they are based on abundant and sound scientific evidence, but laying out that evidence and explaining it would divert most authors from the main point of their arguments, which, in most cases, is to focus on the social and historical aspects of race and racism, not the science.

This book, by contrast, is about the science. The evidence presented here is amassed from the research of hundreds of scientists working in laboratories throughout the world. Most of this research is pure science, conducted without any political or social agenda. I am honored to count myself among these scientists. Colleagues, students, and I have contributed a small portion to the DNA-based information on worldwide human diversity, a sampling of it summarized in this book. Although I am a research geneticist, I believe my greatest qualification for writing this book is an ability to present complex scientific evidence and conclusions in a way that is engaging and understandable to those who do not have an extensive background in the sciences.

A moment ago, I mentioned how, until recently, history books were often sanitized of the atrocities of racism, possibly because those atrocities seem too horrific. This book could potentially suffer criticism for likewise failing to sufficiently recount them. Its focus, however, is *scientific* evidence. That evidence provides a solid foundation that negates both the legacy of historic racism and the pervasive undercurrent of racism that continues to the present. I strongly believe that the history of racism has often been downplayed, and I encourage those who read this book to explore that history as a complement to the scientific information presented here. There are many excellent books, articles, essays, websites, and video documentaries on the subject.[5]

I wish to thank the editors and staff of Prometheus Books for their professionalism and expertise. Although I have attempted to verify all scientific conclusions in this book and support them with reliable evidence, I take full responsibility for any errors. The opinions in this book are mine and do not necessarily represent those of the publisher or my employer. I hope you find this book compelling. Please join me as we explore how science explodes the myth of race.

PROLOGUE

I am writing this paragraph on July 14, *la fête nationale*, also known as Bastille Day in France. On this same date three decades ago, I was riding on a train through Strasbourg, France, colored fireworks lighting up the night sky. The holiday celebrates the storming of the Bastille, a key event in the fall of the French monarchy and the adoption of the *Déclaration des Droits de l'Homme et du Citoyen* (the Declaration of the Rights of Man and of the Citizen), one of the world's most influential documents intended to establish basic human rights, similar to those in the US Constitution. Though the purpose of such documents was to extend those rights to all people, many whose ancestry was not classified as "white" were considered to be inherently inferior by commonly held beliefs and often by law and thus were deprived of these rights.

Today I was on a different train—the number 1 in New York City traveling north from downtown to uptown Manhattan. The ride took about a half hour, and the people I saw coming in and out of the train were diverse, with ancestries from many places. Not only did they appear diverse, many were speaking different languages. I recognized Portuguese and Spanish, languages I speak fluently, and those speaking them had accents typical of Brazil, the Azores, Puerto Rico, Mexico, and Peru. I also heard other languages I did not understand, as well as a wide range of accents in English. The people on the train were probably a mix of local residents, tourists, businesspeople, and students. This sort of vibrant human diversity is now commonplace in major cities throughout the world.

Some celebrate such a mix of human diversity; others deplore it, preferring that so-called races be separated both geographically and reproductively. Even today, some people retain the once-popular belief that the "white" race is superior in intellect, health, and other attributes. Although far more people reject the notion of white supremacy today than in the past, its legacy remains, as evidenced by economic stratification, ongoing segregation, and classification by racial categories. Even among those who reject the supposed superiority of a particular ethnicity over any other, the perception of distinct, genetically determined human races often persists.

This book is, in part, the outgrowth of conflicting race relations I have observed since my childhood. The sounds and images of the civil rights movement were a part of my youth during the 1960s. On television, I watched snippets of Martin Luther King Jr.'s impassioned speeches. The resonant sound of his voice and the hope it evoked left me appalled that people could be treated as inferior simply because of their ancestry. I was eleven years old when I heard the news that he had been assassinated, and I silently grieved for his family.

During my high school years in the mid-1970s, I lived in a small town in eastern Arizona, not far from the border with Mexico. While most of my friends spent the summers bagging groceries for minimum wage, I preferred working outdoors on the local cotton farms alongside undocumented immigrants who had walked across the border to find employment. I often heard people refer to them as if they were less than human, using labels such as "wetbacks" or "spics." I developed friendships with several of them as we walked side by side hoeing weeds in the fields. One man in particular became a close friend. He was obviously well educated and intelligent, and he had abandoned his career as a commercial artist in Mexico City because he could earn more for his family as a migrant farmworker in Arizona. He spoke some English, but most of our conversations consisted of him teaching me Spanish, and I looked forward each day to learning more from him. One morning, he failed to appear, and I later learned that the Border Patrol had taken him away.

He sparked in me an interest in his culture and language that would substantially shape my professional career, eventually culminating in my study of Spanish and Portuguese, a PhD minor in Latin American studies, and decades of research in Mexico, Guatemala, Peru, Bolivia, and Brazil. Although others viewed him as inferior because his skin was darker and his social and political background different, I count him as an influential teacher and friend. Though I cannot recall his name and do not know where he lives now, I will always remember him. I hope he remembers me.

This book is also an outgrowth of what I have discovered as a geneticist from laboratory research on human genetic variation I have conducted with collaborators and from my study of published research from laboratories around the world. Unfortunately, few people are aware of how much is known about the genetic basis of race—or, more accurately, the lack thereof. To many, the notion that race is inherited seems self-evident. Yet extensive genetic research has demonstrated that the genetic variation associated with what most people perceive as race represents a small proportion of overall genetic variation. When viewed on a global scale, there are no discrete genetic boundaries separating so-called races. Rather, the world's human diversity consists of innumerable genetic variations spread throughout the human population in a complex set of multiple overlapping arrays. A proportion is associated with geographic ancestry, but much genetic variation traces its origin to more than one hundred thousand years ago when all humans lived in Africa, and that ancient African variation is now spread throughout the people of the world.

Racial classification is real, but it is based much more on a set of social definitions than on genetic distinctions. Legally defined categories for race differ from one country to another, and they change over time depending largely on the social and political realities of a particular society or nation. The notion of discrete racial categories arose mostly as an artifact of centuries-long immigration history coupled with overriding worldviews that white superiority was inherent—a purported genetic destiny that has no basis in modern science.

Scientific methods that allow for large-scale analysis of human DNA have recently unleashed a flood of genetic information, allowing for detailed analysis of the geographic ancestry of any particular person. Such methods fail to reveal discrete genetic boundaries along traditional racial classification lines. What they do reveal are complex and fascinating ancestral backgrounds that mirror known historical immigration, both ancient and modern. Complex ancestry, rather than simplistic race, is a more accurate and meaningful representation of each person's genetic constitution. This book argues for a scientific approach to unraveling the complexity of human genetic diversity, and against simplistic classification using race as a supposed biological entity. Racial classification must remain social, targeted to meet social realities, to overcome discrimination and provide strong incentives against it in the present and future.

A better understanding of what science tells us about human genetic diversity is of immense importance, particularly because it dispels false notions of what race is. This book summarizes what is known about the genetic basis of human diversity, how it evolved, and what it means. It recounts recent research and how human genetic variation is related to pigmentation, human health, and intelligence, all of which have been attributed to race in the past, often simplistically and erroneously. In the end, we can now read our evolutionary history written in our DNA, and it explains our genetic unity as a species and how our genetic diversity came to be.

CHAPTER 1
WHAT IS RACE?

Mildred Jeter was a woman of African American and Native American ancestry who in 1958 married Richard Loving, a man whose ancestry was European American. Both were citizens of the United States and residents of Virginia. They traveled to Washington, DC, to be married because, at the time, Virginia actively enforced the Racial Integrity Act of 1924, which stated, "If any white person intermarry with a colored person, or any colored person intermarry with a white person, he shall be guilty of a felony."[1] According to this law, the definition of "colored" was derived from the so-called one-drop rule, meaning that any person whose ancestry was not entirely "white"—had even "one drop" of nonwhite blood—was legally considered to be colored. Mildred met Virginia's legal definition of colored, and Richard, the definition of white.

Shortly after their marriage, the Lovings returned to their Virginia home. Late one night, police forcibly entered their house as they slept and arrested them for violating this law. They pleaded guilty in 1959, the judge stating, as quoted in the subsequent Supreme Court case, "Almighty God created the races white, black, yellow, malay and red, and he placed them on separate continents. And, but for the interference with his arrangement, there would be no cause for such marriage. The fact that he separated the races shows that he did not intend for the races to mix."[2]

The case reached the US Supreme Court in 1967, nine years after their marriage, and the court ruled that Virginia's Racial Integrity Act violated the Equal Protection Clause of the US Constitution and was, therefore, unconstitutional. In the unanimous opinion, Chief Justice Earl Warren wrote,

There is patently no legitimate overriding purpose independent of invidious racial discrimination which justifies this classification. The fact that Virginia prohibits only interracial marriages involving white persons demonstrates that the racial classifications must stand on their own justification, as measures designed to maintain White Supremacy. We have consistently denied the constitutionality of measures which restrict the rights of citizens on account of race. There can be no doubt that restricting the freedom to marry solely because of racial classifications violates the central meaning of the Equal Protection Clause.[3]

At the time, sixteen states still had laws prohibiting interracial marriage, known as antimiscegenation laws. As Warren noted in his opinion, the court's decision rendered all of them invalid. Nonetheless, several remained on the books for years after this decision. The final antimiscegenation law in the United States was rescinded by ballot initiative in Alabama in 2000.[4] Similar laws in other countries had persisted late into the twentieth century, notably in South Africa under apartheid. The South African laws—the Prohibition of Mixed Marriages Act and the Immorality Act—were repealed in 1985.[5]

Such laws were based on notions of racial superiority, assumed to be genetically inherited. Few features of humanity are as obvious as the wide array of inherited diversity visible in our outward features. It's also evident that people whose ancestry traces to a particular geographic region typically appear similar to one another and different from those whose ancestries are from other geographic regions. Moreover, we as humans have an almost innate propensity to compartmentalize nearly everything into discrete categories, even when lines that distinguish those categories are complex, blurred, or nonexistent. As an inevitable consequence, people have been subjected to categorization into what we now call human races throughout much of the past several centuries.

The word *race* in English has two distinct meanings derived from two independent origins. *Race* as the act of competitive running derives from the Old Norse word *ras* and is unrelated in both meaning and origin to the word this book addresses. *Race*, as an indicator of a group of people, animals, or plants of common genetic lineage, derives from the Old Italian word *razza* and has been conserved in languages with Latin roots, such as *race* in English and French, *razza* in Italian, *raza* in Spanish, *raça* in Portuguese, and *rasa* in Romanian. Its earliest use in English, in the context of human races, dates to approximately 1500 CE, during the time of the Renaissance.

In modern times, the word *race* is used almost exclusively in reference to humans. We even use it at times to refer to the entire human species, as in *the human race*. But if we go back not too long ago, we find it used more often to denote genetically defined groups of animals and plants. And to understand how the idea of race came to be applied to humans, we need first to examine its earlier usage, especially with domesticated animals.

Until recent times, most people made their living growing crops, tending gardens, and raising animals. Some owned their own land, whereas others were tenant farmers or farmworkers, working on land that belonged to someone else. Whether wealthy or impoverished, most people were familiar with farm animals and were well aware of different types of these animals, which they often referred to as belonging to different races. For example, Charles Darwin uses the term *race* forty-three times in chapter 1 of his best-known book, *On the Origin of Species*, first published in 1859. On a few occasions he writes of race-horses, but in every other instance, his use of the word *race* denotes a genetically distinct group of animals or plants. The following sentence is a case in point: "When we look to the hereditary varieties or races of our domestic animals and plants, and compare them with species closely allied together, we generally perceive in each domestic race, as already remarked, less uniformity of character than in true species."[6]

Interestingly, only three times throughout the entire book does Darwin refer to humans with the word *race*. In one instance, for example, he recognizes the ignorance in his day regarding the cause of human variation: "But we are far too ignorant to speculate on the relative importance of the several known and unknown laws of variation. . . . I might have adduced for this same purpose the differences between the races of man, which are so strongly marked."[7]

In the nineteenth century, *race* denoted a broad grouping of related types of animals, whereas the term *breed* designated a more specific subgrouping, although there was much overlap between these two terms in practical use. For example, bulldogs, poodles, shepherds, hounds, terriers, and retrievers might be considered different races of dogs, whereas distinct subgroups within a race, such as bloodhound, coonhound, basset hound, greyhound, dachshund, and beagle, were (and are) considered different breeds of hounds. In modern usage, the term *race* has mostly disappeared as a designation for animals. The American Kennel Club, for instance, refers to hounds collectively as a group, not a race.[8]

The key distinguishing feature of both breed and race in animals is the ability to breed true, which means that individuals of the same breed or race consistently and predictably produce, generation after generation, offspring of the same type. For example, when two basset hounds mate, all their offspring consistently and predictably have the characteristics of basset hounds, such as long floppy ears, loose sagging skin, short legs, a relatively long body, short hair, and a baying bark typical of hounds. The term *purebred* is used to denote animals certified as members of a particular breed who do not have any recent ancestry from another breed, as documented by well-kept pedigree records. Animal keepers often take great care to ensure that the animals they raise are purebred, and that these animals mate only with members of the same breed, because purebred individuals typically command higher prices at market. A purebred horse with a known pedigree is much more valuable than one without. For dogs, pedigreed purebred puppies command high prices, whereas it is often difficult to find homes for mixed types.

When members of different breeds mate, their first-generation offspring often have a uniform appearance intermediate between their parents for some traits but more closely resemble one parent for other traits. These offspring are known biologically as *hybrids*, and when two hybrids mate, they often do not breed true. Instead, their offspring tend to vary in the traits they display. Darwin noted this phenomenon in the *Origin of Species* when he wrote, "The slight degree of variability in hybrids from the first cross or in the first generation, in contrast with their extreme variability in the succeeding generations, is a curious fact and deserves attention."[9] He recounted in this passage a phenomenon that animal and plant breeders had known for centuries: the uniformity of hybrids and the variability of their offspring, or, in other words, the inability of hybrids to breed true.

It is no surprise that throughout the past several centuries, people have used the term *race* to describe groups of people in much the same way it was used in past centuries to describe groups of animals. People with ancestry from a particular region of the world tend to share certain inherited physical characteristics, such as similar facial features and eye, hair, and skin pigmentation. The children of parents with similar regional ancestry typically inherit similar features, resembling their parents. However, the children of parents with substantially different ancestral backgrounds often have an appearance that is intermediate between that of their two parents. And in subsequent generations, the offspring may vary.

In part because of the obvious similarities between animals and humans for how traits are inherited, and in part because of cultural, political, and religious traditions, notions of racial purity and superiority have surged and ebbed yet persisted, crossing the boundaries of culture, geography, politics, and time. They are still with us today, and some of the most insidious actions based on notions of racial supremacy happened not long ago. They were the foundation of two related movements that became codified into law in various places and times: the eugenics movement and antimiscegenation.

In 1883, Francis Galton coined the term *eugenics* to denote the intentional genetic improvement of humans, much like the intentional breeding of animals and plants. The idea of direct selective breeding of humans was appalling to most, but discouraging or legally preventing people from procreating who were considered genetically unfit seemed much more acceptable. The eugenics movement thrived on a universal undercurrent of the supposed inherent superiority of the "white race"—people whose ancestry was northern European, British, or Scandinavian. Books and articles endorsing eugenics proliferated. For instance, Reginald Punnett, a famous British geneticist, wrote in 1913, "By regulating their marriages, by encouraging the desirable to come together, and by keeping the undesirable apart we could go far towards ridding the world of the squalor and the misery that come through disease and weakness and vice."[10]

The first eugenics laws in the United States requiring mandatory sterilization of people considered unfit to reproduce were implemented in 1907, and between 1910 and 1930, many US states and European countries instituted laws mandating involuntary sterilization of people in prisons and mental institutions. Along with mandatory sterilization laws came antimiscegenation laws prohibiting interracial marriages, usually defined as marriages between "whites and nonwhites" or "whites and coloreds," in an attempt to maintain the purity of the so-called white race. Under the guise of eugenic improvement and racial purity, and what the implementation of eugenic measures could supposedly do to improve human society, notions of racial superiority continued in popularity, but with a purported scientific foundation.

The worldwide popularity of the eugenics movement expanded during the first decades of the twentieth century, reaching its climax under the Nazi regime in the 1930s and '40s. A fundamental tenet of Nazism was the superiority of the so-called Nordic class of the Aryan master race, typified by light skin and hair, blue or green eyes, and Germanic origins, presumed to be associated with superior intellect and physical prowess. The Nazi regime exterminated millions of people because their genetic background was classified as subhuman (*Untermenschen*, literally "under-men") and they were deemed a threat to the Aryan race. The majority of those who were killed were European Jews, approximately six million, although Roma (gypsies) and Slavic people, mostly Polish and Russian, were also killed, as well as people designated as homosexual and those with mental disabilities.[11]

Lesser known, but likewise intended to preserve and increase genetic integrity and superiority, were the Nazi antimiscegenation laws and the Lebensborn program. The antimiscegenation laws were part of the Nuremberg laws, which initially were anti-Semitic, defining people with four German grandparents as German, those with three or four Jewish grandparents as Jewish, and those with one or two Jewish grandparents as *Mischling*, meaning mixed or half-breed. The laws were complex but essentially required racial classification of people and discriminated against those considered non-Aryan. These laws were eventually extended to discriminate against gypsies and black Africans as unsuitable for marriage with a person defined as German.

The *Lebensborn* (wellspring of life) program was a secretive Nazi effort to develop a superior human race through direct and intentional human breeding. Instituted at the same time as the Nuremberg laws in 1935 and managed by Heinrich Himmler, the program aimed to identify women and men considered to be racially pure Nordic Aryans—mostly young women who applied to the program and passed the racial superiority screening test and SS soldiers who served as the biological male parents. Himmler decreed that SS soldiers should father as many *Lebensborn* children as possible; marriage was not required. The children were to be raised in *Lebensborn* homes, furnished with loot from Jewish homes. As many as 17,500 children were born under the program, most placed with adoptive families after the war.[12]

As these secret actions were uncovered after World War II, eugenics laws fell into disrepute and were eventually repealed or left unenforced. By then, however, more than sixty thousand people had been involuntarily sterilized in the United States under such laws.[13] Although some antimiscegenation laws were repealed, many remained in force well after the end of World War II, a result of belief that racial purity should be maintained. In the United States, the legality of those laws came to a dramatic and definitive end in the Supreme Court case *Loving v. Virginia*, recounted at the beginning of this chapter.

Antimiscegenation and eugenics laws stem from the notion that superior racial purity should be preserved, similar to maintenance of genetic purity in breeds of domesticated animals. Although similar patterns of inheritance are evident in animals and humans, there are some crucial flaws in assuming that breeds of animals equate to so-called human races. First, humans have intentionally bred domesticated animals to achieve the true-breeding characteristics that purebred animals have, resulting in extremes of variability among breeds—and uniformity within them—that vastly exceed the extremes of natural variation in humans. In dogs, for example, Chihuahuas and Great Danes differ by extremes in size, variation that far outweighs the outward variation among humans, and those extremes are a direct result of intensive breeding. Although restrictions on mating have been a part of various human cultures for most of history, humans have never been subjected to the type of intense selection for extreme types that is typical in animal breeding. Geographic proximity and cultural traditions—not intentional breeding—have historically been the most powerful factors influencing choice of mates in humans.

A second flaw in equating human races to animal breeds is that, historically and especially in modern times, humans have been a highly mobile species. Although some degree of historic reproductive isolation is responsible for part of the geographic distribution of genetic diversity in humans, there is evidence in our DNA that major and complex human migrations have dispersed, and continue to disperse, much of the genetic diversity in our species.[14]

These and other factors have ensured that the lines separating the worldwide geographic groups of humans are so blurred they are impossible to define. Harvard professor and geneticist Richard Lewontin describes his perception of the situation quite succinctly:

> Racial classification is an attempt to codify what appear to be obvious nodalities in the distribution of human morphological and cultural traits. The difficulty, however, is that despite the undoubted existence of such nodes in the taxonomic space, populations are sprinkled between the nodes so that boundary lines must be arbitrary.[15]

The article by Lewontin from which this quote is extracted bears the title "The Apportionment of Human Diversity," and it is one of the most foundational and misconstrued articles ever published on the biological basis of race in humans. Before the 1960s and '70s, scientists had few methods available to confidently examine genetic diversity among people. They based estimates of diversity largely on appearance, making assumptions about how much diversity was inherited and how much was a consequence of nongenetic, often termed environmental, variation. Moreover, much of the assessment of human diversity was biased, as Lewontin puts it, by "those characters to which human perceptions are most finely tuned (nose, lip, and eye shapes, skin color, hair form and quantity)."[16] Variation for such traits represents only a small subset of overall human genetic diversity, and they are used for classification largely because they are the varying characteristics our brains are attuned to most readily recognize—and those we consciously or unconsciously associate with the geographic ancestries of different people. Other outward characteristics that vary among people do not fall neatly within traditional racial categories.

Adult body height, for example, is genetically determined to a large extent, and it varies considerably among people from all regions of the world. If we were to segregate people by height, there would be little association along traditional lines of racial classification. The tallest people in the world, on average, are the Dutch, and among the tallest are the Masai of East Africa.

Several indigenous groups living in tropical regions of the world are substantially shorter on average, including the Mbenga, Mbuti, and Twa of Africa; the Andamanese, Aeta, Batak, Rampasasa, Semang, and Taron of south and Southeast Asia; the Djabugay of Australia; and the Yąnomamö of Amazonian South America. The genetic basis for some of these extremes in stature has been well documented and is likely a consequence of natural selection.[17] Clearly, these variations in height do not mirror traditional racial or geographic classification schemes.

As another obvious example, pattern baldness is a common inherited trait that varies among people (mostly in men and to a lesser extent in women) whose ancestries trace to various parts of the world. Although readily visible, and less prevalent in certain regions of the world than in others, variation for pattern baldness has never been a criterion for racial classification. It makes no sense to lump those with pattern baldness into one racial classification and those without it into another. Only a subset of genetically determined traits that vary in humans is correlated with geographic regions of ancestry.

Other inherited characteristics that are not outwardly apparent, such as blood types and other biochemical traits, vary among people. It is, of course, impossible to tell simply by looking at someone what her or his blood type is, but it is possible to precisely identify blood types and other inherited biochemical traits through laboratory analysis. By the 1970s, scientists had developed methods that allowed them to quantify some of this unseen genetic diversity with exceptionally high accuracy. Lewontin asked whether such biochemical diversity was correlated with traditional racial classifications. In other words, were the physical variations people tend to associate with different races borne out in the precise genetic variations scientists could measure in the laboratory?

His conclusion was shocking—unexpected in its magnitude—and has shaken the foundation of racial classification ever since its publication. Lewontin examined data from seventeen different genes in people classified into seven groups by geographic origin, as named in his article: Amerinds, Australian Aborigines, Black Africans, Caucasians, Mongoloids, Oceanians, and South Asian Aborigines. He found that the degree of genetic variation *within* each group exceeded that *between* different groups. In other words, people within a particular racial group varied more among themselves than their overall group varied from other groups. On average, according to his calculation, 85 percent of overall genetic diversity fell within groups, whereas only 15 percent could be attributed to between-group differences. He also noted that "the difference between populations within a race accounts for an additional 8.3%, so that only 6.3% is accounted for by racial classification."[18]

As an example, A, B, AB, and O blood types vary within the groups identified by Lewontin, albeit not in the same proportions within each group. For instance, all four types are present within the European group and the African group, with type O being the most prevalent within both groups, albeit more prevalent within the African group. Because all four types are present within both groups, a person whose ancestry is European might be incompatible as a potential blood donor for another person with European ancestry, whereas a person with African ancestry may be a compatible donor. In fact, blood banks in hospitals currently identify blood based only on biochemical blood typing, with no reference to racial identification of donors (although this has not always been the case).[19]

Lewontin's analysis rapidly gained popularity, for it seemed to provide a scientific case against a biological basis for racial inequality, and the political timing was right. When Lewontin's article was published in 1972, the civil rights movement in the United States had by then gained considerable political support and was a frequent issue in the news. Lewontin's article was often cited then, as it is now, as evidence that there is scant scientific basis for racial classification. Such classification was said to be more social than biological.

Thus, it was quite a turnaround in 2003 when Cambridge University professor A. W. F. Edwards, one of the world's foremost statisticians and population geneticists, published a counter-article titled "Human Genetic Diversity: Lewontin's Fallacy." Edwards makes it clear that "there is nothing wrong with Lewontin's statistical analysis of variation, only with the belief that it is relevant to classification."[20] Edwards points out that Lewontin examined each of seventeen genes independently, but that an assumption of independence was unwarranted; variation for one gene may be correlated with variation for another, and such correlations must be taken into account when using such data for classification. As a simple example, Edwards points to a common practice in anthropology of measuring the length (back to front) and breadth (side to side) of skulls. It is possible to focus on length and breadth independently and determine variation for each among human skulls without reference to the other. But doing so misses the important point that the two are correlated; an increase in overall skull size results in a concomitant increase in *both* length and breadth. Edwards then highlights a number of studies of human genetic diversity that take such correlation into account. The correlation is essential for describing human genetic diversity and systems intended to classify such diversity.

Challenging a number of published statements based on Lewontin's research, Edwards emphatically concludes,

> It is not true [as Lewontin claimed] that "racial classification is . . . of virtually no genetic or taxonomic significance." It is not true, as *Nature* claimed, that "two random individuals from any one group are almost as different as any two random individuals from the entire world," and it is not true, as the *New Scientist* claimed, that "two individuals are different because they are individuals, not because they belong to different races" and that "you can't predict someone's race by their genes." Such statements might only be true if all the characters studied were independent, which they are not.[21]

So who is right? When Lewontin did his analysis in 1972, the number of people and the number of genes he examined were just a minuscule fraction of the number of people and genes that have been studied today. More extensive research has confirmed the conclusions of *both* Lewontin and Edwards—negating, however, many of the claims that others inferred from Lewontin's results. In 2004, human geneticists Lynn Jorde and Stephen Wooding of the University of Utah School of Medicine summarized the results from several large-scale studies. First, they confirmed that humans as a species are much less diverse than many other species. According to their estimates, people worldwide differ on average by about 0.1 percent, evidence that all humans are genetically quite similar to one another. They then confirmed Lewontin's major conclusion:

> Of the 0.1% of DNA that varies among individuals, what proportion varies among the main populations? Consider an apportionment of Old World

populations into three continents (Africa, Asia, and Europe), a grouping that corresponds to a common view of three of the "major races." Approximately 85–90% of genetic variation is found within these continental groups, and only an additional 10–15% of the variation is found between them. . . . These estimates . . . tell us that humans vary only slightly at the DNA level and that only a small proportion of this variation separates continental populations.[22]

The accumulation of information from many individuals and their DNA also revealed correlations that correspond with the three major continents of geographic ancestry. In all cases, Jorde and Wooding were able to accurately assign native Europeans, east Asians, and sub-Saharan Africans (Africans whose ancestral origin is south of the Sahara) to their respective continents of origin even when the samples were examined for DNA variation alone without taking into account any other characteristic.

The authors were quick to point out, however, that their data *do not* support the traditional boundaries of race:

> [I]t might be tempting to conclude that genetic data verify traditional concepts about races. But the individuals used in these analyses originated in three geographically discontinuous regions: Europe, sub-Saharan Africa, and east Asia. When a sample of South Indians, who occupy an intermediate geographic position, is added to the analysis, considerable overlap is seen among these individuals and both the east Asian and European samples, probably as a result of numerous migrations from various parts of Eurasia into India during the past 10,000 years. Thus the South Indian individuals do not fall neatly into one of the categories usually conceived as a "race." . . . Ancestry, then, is a more subtle and complex description of an individual's genetic makeup than is race.[23]

This quotation makes a crucial point regarding sampling and interpretation of scientific data from human populations, one that often leads to misconceptions regarding racial classification. Our species is highly mobile; people have been migrating throughout the world for thousands of years, resulting in a distribution of genetic variation that is complex, overlapping, and more continuous than discrete. When people are sampled from discontinuous geographic extremes (such as northern Europe, east Asia, sub-Saharan Africa, Australia, and the Americas), without including samples from people whose ancestry originates between these extremes (such as the Middle East, central Asia, and south Asia), data from these discontinuous samples, not surprisingly, portray a discontinuous distribution and can be classified into discrete groups. Even so, about 85 percent of genetic variation is still shared among these so-called discrete groups. A more accurate representation of the worldwide human population, however, is a sampling of people whose ancestral origins are spread throughout the world. When DNA is analyzed from a more geographically broad sample of people, substantial overlap is evident, and the notion of discrete racial boundaries disappears.

The massive amount of scientific evidence we currently have reveals complex and interwoven histories of human diversity, which provide a far more compelling case against racism than ever before. In the upcoming chapters we explore this evidence and what it tells us about the history of our worldwide human family.

CHAPTER 2
AFRICAN ORIGINS

On the ceiling of the Sistine Chapel in the Vatican are Michelangelo's magnificent fresco paintings, one of the greatest triumphs in the history of art. They depict biblical stories, the three central panels portraying the creation of humankind. Perhaps the most famous is a powerful image of God surrounded by angels and reaching out to touch Adam's extended finger, granting him the spark of life. The center panel shows Eve emerging from Adam's rib cage—Adam having been divinely anesthetized into a deep sleep—while God beckons Eve to come forth. The third panel shows two stories, one with Lucifer as a humanized serpent tempting Adam and Eve, and the other, the expulsion of Adam and Eve from the Garden of Eden by a sword-wielding angel robed in crimson (figure 2.1).

Figure 2.1. Fresco paintings on the Sistine Chapel ceiling by Michelangelo. *Image from* Michelangelo: des meisters werke in 166 Abbildungen *by Fritz Knapp, 1906.*

Congruent with widely accepted beliefs in sixteenth-century Italy about the biblical account of human origins, Michelangelo depicted Adam, Eve, God, Lucifer, and the angels with Caucasian features. The same is true for innumerable depictions of Eve and Adam by European and American artists who lived before and after Michelangelo. The trend persists to the present. The Creation Museum in Petersburg, Kentucky—a suburb of Cincinnati, Ohio—displays life-size mannequins of Adam and Eve with Caucasian features, consistent with most contemporary American religious depictions of Eve and Adam.

Modern science paints a dramatically different picture of human origins and the genetic basis that underlies worldwide human diversity. The scientific picture is founded on abundant evidence from a wide range of sources, including anthropological excavations, research on how human genetic characteristics are distributed and inherited, and large-scale analysis of DNA from thousands of people representing the world's indigenous populations. This evidence reveals in abundant and exquisite detail where and when our species originated and how humans populated the world.

The evidence can be divided into two major categories. The first is anthropological evidence: skeletal remains of ancient humans unearthed at archaeological sites and relics of human activity such as tools, ornaments, potsherds, waste, and building remnants. The second is genetic, which includes the distribution of inherited characteristics among humans, coupled with a flood of recent evidence from DNA that is increasing at an exponential rate. Although these two major lines of evidence have been mostly independent, they converge to tell essentially the same story. The major conclusion, now supported by overwhelming evidence from both anthropology and genetics, is solid and no longer questioned by scientists: *the human species originated in Africa.*

The anthropological evidence alone is powerful. The oldest remains of what anthropologists call "anatomically modern humans" (skeletons with features that resemble those in modern humans) are exclusively from Africa, dating to about two hundred thousand years ago. By contrast, the earliest remains of anatomically modern humans outside of Africa thus far discovered are about one hundred thousand years old. These are from an ancient population that lived in what is now Israel, determined from human remains found in the Qafzeh and Es Skhul caves.[1] This population apparently suffered extinction around seventy thousand years ago, leaving no modern descendants. Remains of anatomically modern humans dating to sixty thousand years ago or less are common both inside and outside of Africa, and the more recent they are, the more abundant and widespread they tend to be. The locations and dates of these remains are consistent with a scenario in which humans emigrated out of Africa about sixty thousand to seventy thousand years ago, and their descendants, through many generations, eventually populated the rest of the world.

There is, however, evidence that other humanlike species lived in Asia and Europe long before the dates assigned to remains of anatomically modern humans in these regions. *Homo erectus*, for example, emigrated out of Africa at least two million years ago, before anatomically modern humans existed, migrating across Asia into what is now China and Southeast Asia. Neanderthals evolved in Europe from ancient ancestors who apparently emigrated about seven hundred thousand years ago from Africa into Spain across what are now the Straits of Gibraltar. Neanderthals then spread throughout Europe and much of the Middle East.[2]

The prevalence of fossils outside of Africa from humanlike species that predate anatomically modern humans led some anthropologists to propose what is known as the *multiple-origins hypothesis*, the idea that modern humans arose in several places throughout the world from humanlike species already there. The alternative is the *single-origin hypothesis*—sometimes called the "Out of Africa" hypothesis because it proposes that anatomically modern humans evolved in sub-Saharan Africa, and then a subset of these humans later emigrated out of Africa and became the ancestors of everyone whose ancestry is not confined to Africa. Notably, both hypotheses propose a sub-Saharan African origin for the ancient ancestors of all modern humans.[3]

As anthropological evidence from numerous excavations accumulated over time, it tended to favor the single-origin hypothesis. For instance, anatomically modern humans and Neanderthals overlapped in Europe and the Middle East from about sixty thousand years ago, when people immigrated into the Middle East from Africa, to twenty-six thousand years ago, when Neanderthals suffered extinction. Had humans originated from Neanderthals in those regions, we would expect to observe fossils that gradually change from Neanderthal anatomy to modern human anatomy. Instead, both Neanderthal and modern-human fossils simultaneously date to this time period, consistent with a scenario in which modern humans immigrated into regions already occupied by Neanderthals.

Once Neanderthal DNA was extracted and examined, it became evident that for people throughout the world, except those whose ancestry is entirely African, a small amount of DNA (less than 4 percent) is derived from Neanderthals.[4] For instance, I have 2.8 percent Neanderthal DNA, according to a DNA test. Some have argued that this observation supports the multiple-origins hypothesis. On the contrary, it supports the single-origin hypothesis with limited mating between modern humans and Neanderthals in the Middle East and Europe from about sixty thousand to twenty-six thousand years ago, when the geographic distribution of the two overlapped. If this is the case, we expect to see no Neanderthal DNA in people of entirely sub-Saharan African ancestry because, having evolved in Europe and the Middle East, Neanderthals were not present in Africa at the time when modern humans emerged in Africa as a distinct species. This is exactly what the data show: people entirely of sub-Saharan African ancestry have no Neanderthal DNA.

This discovery is just one among a deluge of discoveries in recent years from DNA analysis. The most abundant, detailed, and compelling evidence of our origin and history is genetic—from DNA—and it indisputably supports the single-origin hypothesis. This evidence reveals the ancient history of our species, including how and when humans populated different regions of the world, and, in doing so, it explains much about the concept of race in humans. It is here—in the patterns of diversity in our DNA—that we find the strongest evidence of an African origin for modern humans, evidence that, for the most part, has been collected and analyzed independent of the evidence from anthropology. Yet the DNA evidence is entirely congruent with the anthropological evidence.

Our genetic history is written in our DNA. The word *written*, of course, is metaphorical; no one took a pen and wrote out our genetic information. But it is an apt metaphor because DNA carries linear hereditary information, much like the linear arrangement of letters in written language. And scientists can now readily decipher and read that information.

Each DNA molecule is long and slender, composed of *bases*, analogous to the letters on the page of a book. The number of bases in each DNA molecule in humans ranges from the thousands to more than two hundred million. The English alphabet has twenty-six different letters. The DNA alphabet is much simpler, with only four different bases, represented as the letters T, C, A, and G. Because of the linear arrangement of DNA information, we can write the sequence of any DNA molecule with these four letters. For example, a very small part of the DNA responsible for determining the amount of pigment in human eyes, hair, and skin has this sequence:

AGCATCCGGGCCTCCCTGCAGC

DNA molecules are typically double stranded, which means there are two parallel strands of bases in each molecule. Each base in one strand is paired with a corresponding base in the other, and they pair according to strict rules: T pairs with A, and C with G. So we can write the sequence we just examined as a double-stranded sequence by matching the bases according to the T–A and C–G rules, with the bases in one strand directly below their paired bases in the other:

AGCATCCGGGCCTCCCTGCAGC
TCGTAGGCCCGGAGGGACGTCG

Notice that every T is paired with an A and every C with a G. This pairing of bases along the two strands makes possible one of DNA's essential functions: its ability to faithfully replicate. When a DNA molecule is replicated, it splits into single strands, and a process in your cells makes new strands paired to the old ones according to the T–A, C–G pairing rules. When replication is complete, the two resulting DNA molecules are identical in their sequences, each with an original old strand and a newly made strand:

AGCATCCGGGCCTCCCTGCAGC old strand
TCGTAGGCCCGGAGGGACGTCG new strand

AGCATCCGGGCCTCCCTGCAGC new strand
TCGTAGGCCCGGAGGGACGTCG old strand

This precise replication is the foundation for biological reproduction. Ultimately, the fact that children genetically and outwardly resemble their parents is, at its most fundamental level, a consequence of DNA replication and its fidelity.

Although DNA usually replicates faithfully, producing two identical double-stranded molecules from a single original double-stranded molecule, in rare instances the sequence changes ever so slightly, often by just one base pair. Once changed, the altered sequence is then replicated faithfully from that point forward; in other words, the change is inherited. When these changes initially occur, they are called *mutations*. However, scientists often instead refer to these inherited changes as *variants* because most of the variation in our DNA is considered normal—a consequence of mutations that happened long ago in our ancestors, variants that have since been inherited over many generations. All the variants in the DNA of all people constitute the genetic diversity of our species, and these variants originated as millions of mutations that arose at various times in our ancestors.

Many of these variants have no effect whatsoever. Others may influence the variation in our outward characteristics. For example, a variant associated with differences in skin, hair, and eye pigmentation in humans happens to be within the DNA sequence we just examined. Some people have the sequence

AGCATC**C**GGGCCTCCCTGCAGC
TCGTAG**G**CCCGGAGGGACGTCG

whereas others have the sequence

AGCATC**T**GGGCCTCCCTGCAGC
TCGTAG**A**CCCGGAGGGACGTCG

And some people have *both* of these sequences, one inherited from each parent. Notice that the boxed C–G pair in one sequence is a T–A pair in the other sequence. The C–G pair, as it turns out, is the original version, and the T–A variant arose from it long ago through mutation. We typically use the term *ancestral variant* to denote the original DNA sequence carried in ancient humans and the term *derived variant* for any variant that arose by mutation from the ancestral variant. In this example, the T–A variant is derived, and it is associated with less pigmentation in eyes, skin, and hair. It is highly prevalent in people with northern European ancestry, whereas the ancestral C–G pair is associated with a greater amount of pigment and is most common in people whose ancestries trace to parts of the world outside Europe.

Having briefly discussed the nature of DNA and its variants, we're ready to examine the evidence of our origins and see what it tells us about human diversity. Every human being has about six billion pairs of bases in the forty-six DNA molecules we carry in each of our cells. And we all are extremely similar in the base pairs we carry, about 99.9 percent identical on average. The tiny proportion of DNA that makes each of us genetically different from everyone else, however, is significant. Because 0.1 percent of six billion is six million, the number of variants one person carries relative to another can number in the millions, and the degree of difference depends to some extent on how closely related those two people are.

The differences we carry in our DNA encode the genetic diversity of humans throughout the world. And the measure of diversity is not just the number of variants but also their prevalence. A variant that is present in less than 1 percent of people contributes less diversity than one that is present in 10 percent of people. As scientists have studied variants and their prevalence, one major conclusion has emerged in essentially every large-scale worldwide study: the highest diversity by far is among people whose recent ancestry is African.

I use the word *recent* here because if we go back far enough, everyone's ancestry is African. In this case, *recent* means within the past several thousand years. And *African*, in this sense, means sub-Saharan African. It does not include most people who currently live in northern Africa (mostly in Egypt, Libya, Algeria, and Morocco), who descend, to a large extent, from immigrants who entered northern Africa from the Middle East, the Balkans, and Europe. Nor does it include Africans who descend predominantly from people who immigrated to Africa from Europe during the past several centuries, such as South Africans descended from Dutch and British immigrants.

The reason people with entirely African ancestry have the highest diversity in the world is straightforward and can be illustrated with a simple analogy. When I was in grade school, my friends and I used to play games with marbles, and I had a large collection of them. They were highly diverse, with a multitude of colors and patterns. Some marbles in my collection were identical to others, especially those with solid colors—and I had numerous copies of those, lots of solid red, green, blue, orange, yellow, black, and white marbles. Several were much more varied in their coloration but also more rare, represented just once or a few times in my collection. Now, imagine thousands of these marbles, both common and rare types, are mixed randomly in a large container. Imagine reaching into this large container with a cup and scooping out about fifty marbles. The overall diversity of marbles in the cup is not likely to be the same as in the container. Some of the rare types will almost certainly be absent from the cup, remaining only in the large container. Most, and perhaps all, of the more common types are likely to be in the cup, albeit in somewhat different proportions than in the container. In any case, the collection of marbles in the cup is likely to be less diverse than the one in the large container, and the reason has to do with sampling a less diverse subset of individuals from a much larger and more diverse collection.

This same sampling phenomenon happens genetically when a group of people emigrates from a region. Those who leave the original population constitute a subset of that population, and they carry a subset of the overall genetic diversity of the original population. They become the founders of a new population containing the more limited genetic diversity of the emigrants. Now, imagine that several generations later, another group of people emigrates away from the descendants of the first group of emigrants. The diversity diminishes even further in these secondary emigrants. Each group of subsequent emigrants carries a less diverse subset of the diversity that was present in the population from which they originated. Thus, the greatest diversity should be among people whose ancestors constituted the original population, and the region where they live typically represents the region of origin. For humans, that region unquestionably is sub-Saharan Africa.

Studies of human genetic diversity consistently show the diversity in people of African ancestry is the highest in the world. And the evidence from both anthropology and DNA strongly supports a scenario in which people emigrated out of Africa about sixty thousand to seventy thousand years ago and founded what ultimately became the rest of the world's human population, carrying with them a subset of the diversity in Africa. Although far more people in the world's current human population are descended from these out-of-Africa emigrants than from people who remained in Africa, the majority of the world's genetic diversity is still indigenous African.

This observation explains and augments the major conclusion of Lewontin's 1972 study, discussed in the previous chapter. His observation that there is more diversity within major geographic groups than among them is largely a result of the original diversity that was present in Africa more than one hundred thousand years ago, when all humans lived there. The emigrants who left Africa carried a subset of that original diversity in their DNA, and, as a result, many of the same variants are present in people throughout the world, both African and non-African. Thus, much of the variation within major groupings of people is original African variation predating the out-of-Africa diaspora.[5] More recent variants—those that originated after the dispersal of humans throughout the rest of the world—should be more rare and concentrated in geographically localized populations. The more recent these variants are, the more rare and geographically localized they should be. And, consistent with Edwards' description of correlation, these more recent variants tend to be correlated with one another according to the region of more recent geographic origin.[6]

Patterns of genetic diversity are evident in all types of DNA, but they have been most extensively documented in what is called *mitochondrial DNA*, and for some very good reasons. In general, each of us inherits about half our DNA from our mother and half from our father. But mitochondrial DNA is a very important exception. It resides in different compartments of our cells than the rest of our DNA, compartments called mitochondria, and each of us inherits our mitochondrial DNA *exclusively from our mothers*. Thus, variants in mitochondrial DNA are inherited purely through the maternal lineage, from a mother to all her children, but transmitted to the following generation only through females—from mother to daughter. Although males have mitochondrial DNA, it is a hereditary dead end; they do not pass it on to their offspring.[7]

Mitochondrial DNA is relatively small compared to the rest of our DNA—only 16,569 base pairs, compared to slightly more than six billion base pairs in the rest of our DNA (slightly more than three billion inherited from each parent). Thus, it is relatively easy for scientists to track and sequence it. However, it has one feature that makes it especially useful for studying diversity: mitochondrial DNA does not recombine.

For most of your DNA, you inherited half from your mother and half from your father. Go back a generation, and each of them inherited essentially half of their DNA from each of their parents, so about one quarter of your DNA is from each of your four grandparents. During the formation of an egg cell in your mother, the DNA molecules from her parents came together and exchanged segments, shuffling the information they carried. This same type of shuffling also happens during the development of sperm cells in males. And this shuffling recombines maternal and paternal DNA in every generation.

Mitochondrial DNA, however, does not recombine. It is replicated faithfully each generation and inherited through multiple generations exclusively through maternal lineages. If a mutation happens in mitochondrial DNA, it may end up being transmitted as a variant from mother to daughter through subsequent generations. Then, in a later generation, a new variant may originate against the background of the first variant. Some people inherit just the first variant, and others inherit the second variant on the background of the first. This pattern then repeats itself for additional variants that arise at various times and places through many generations. Because there is no recombination, new variants are superimposed on backgrounds of previous variants.

A simple analogy illustrates how these various layers of variants that originated as mutations at different times in mitochondrial DNA allow scientists to reconstruct ancient human genetic history. Before the printing press became available, scribes made handwritten copies of valuable manuscripts. In most cases, the original manuscript had been lost or was not available, so scribes made copies from other copies. Occasionally, a scribe made an error—perhaps a word copied incorrectly or left out—and other scribes subsequently copied the change. Then, later, another scribe made yet another error, adding it to a manuscript with the previous error that had persisted through several rounds of copying, so now two errors were present. Later, another scribe added yet another error to these two. Over time, errors accumulated, more recent ones added to earlier ones. The earlier errors tend to be more widespread, whereas the more recent errors are localized among fewer copies. Modern literary scholars can compare all existing copies of a particular work and hierarchically group them, ultimately extrapolating back to determine much of the original wording.

Scientists who examine the sequences from mitochondrial DNA can reconstruct the same sorts of hierarchical groupings based on the variants they find. Widespread variants in large numbers of people from diverse geographic origins must be the most ancient. Rarer variants in smaller groups of people with a more limited geographic origin must be more recent. And, in each case, newer sets of variants are superimposed on identifiable sets of older variants, allowing scientists to classify different sequences of human mitochondrial DNA hierarchically into numerous small groups clustered within sets of larger groups, and yet again within sets of even larger groups. Ultimately, the variants coalesce into a single group of the most ancient variants, which is the trunk of the maternal human family tree. Furthermore, many of the mitochondrial DNAs examined are from people who belong to indigenous populations, groups of people who have been geographically and reproductively isolated for many generations. By comparing mitochondrial DNA sequences with the geographic regions where these indigenous people live, scientists can reconstruct the geographic migration patterns of ancient humans.

Our current understanding of mitochondrial DNA diversity is extensive, highly reliable, and derived from complete mitochondrial DNA sequences from thousands of indigenous people from various parts of the world. Moreover, Neanderthal mitochondrial DNA sequences have been obtained from the remains of several individuals, allowing for comparison of modern human and Neanderthal mitochondrial DNA. In an earlier book I authored, *Evolving: The Human Effect and Why It Matters*, I devoted most of a chapter to the details of how mitochondrial DNA evidence reveals ancient human emigrations. Here, we'll focus on a few of the major conclusions.

First, the highest mitochondrial DNA diversity in the world is found in people whose maternal ancestry is African.[8] These highly diverse African mitochondrial DNAs can be classified into several large and diverse ancient groups, called mitochondrial *haplogroups*. Seven major haplogroups are found in people whose maternal ancestry is African: haplogroups L0, L1, L2, L3, L4,

L5, and L6. L0 is the most distinct, and it diverged from the ancestral type for all the others very early in Africa, probably more than 190,000 years ago. People who carry the L0 haplogroup belong mostly to indigenous African groups who speak Khoisan languages and reside predominantly in the south of Africa. The remaining haplogroups diverged at various times from a common ancestral type and are found in Africa, largely among people who speak other African languages.

Although each haplogroup has its own intriguing history, the L2 and L3 haplogroups are two of the most relevant in terms of emigration out of Africa, both ancient and modern. People carrying L2 in ancient times (beginning about ninety thousand years ago) immigrated to the western regions of Africa but later spread throughout Africa. Today, L2 is the most prevalent African haplogroup among people whose recent maternal ancestry is African, both within and outside of Africa, including people whose maternal ancestors were taken from west Africa as slaves. For instance, most people who identify themselves as African American (in the North and South American continents and Caribbean islands) carry the L2 haplogroup.

People carrying haplogroup L3 spread anciently throughout the northern parts of Africa. It is here that we find powerful evidence supporting the single-origin hypothesis for humans whose ancestry lies outside Africa. All non-African mitochondrial haplogroups, from all people anywhere in the world outside Africa, trace their origin to L3. Because this observation points to a single African origin for all non-African mitochondrial haplogroups, it contradicts the multiple-origins hypothesis. It points to a group of people (or perhaps several groups) who carried L3 and emigrated out of northeastern Africa about sixty thousand to seventy thousand years ago. Their descendants immigrated into west-central Asia and became the genetic founders of people who populated the rest of the world in ancient times.

Newer variants arose against the L3 background in people who lived in the Middle East and had descended from ancestors who left Africa, causing L3 to diverge into two major haplogroups, called M and N. A third major haplogroup, called R, then diverged from N. Even though they originated from the African haplogroup L3, we'll refer to M, N, and R as *non-African* from this point forward because the variants that define them arose outside Africa and were the founding haplogroups for all other haplogroups that arose outside Africa. The map in figure 2.2 is a simplified version showing the major routes of migration and the mitochondrial haplogroups.

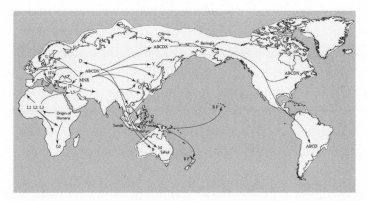

Figure 2.2. Major migration routes for early human diasporas, as revealed by mitochondrial DNA analysis. The area in white represents the approximate ancient landmasses and ice sheets when sea levels were lower during the last major ice age, when these migrations took place. The modern continents are outlined in black.

As an example, let's focus on people whose maternal ancestry is Native American—from the northern regions of Alaska and Canada to the southern tip of South America. As shown in figure 2.2, they carry five mitochondrial haplogroups, called A, B, C, D, and X. The first four (A, B, C, and D) are prevalent and widespread throughout the Americas. A relatively small number of people with ancient North American ancestry carry the haplogroup X, which is always rare wherever it is found. All five of these haplogroups are also present in Asia, and the patterns of variants in Native Americans make it clear that people carrying them emigrated from Asia to North America about fifteen thousand years ago, probably in more than one emigration event, across a land bridge called Beringia that connected what are now northeastern Siberia and Alaska.[9] Ancient emigrations from Asia to the Americas ceased when the land bridge was inundated by rising sea levels at the end of the most recent ice age, about eleven thousand years ago.[10] Thus, the ancient ancestry of Native

Americans is undoubtedly Asian. Interestingly, these five haplogroups that are present in people from both Asia and the Americas arose from all three major non-African haplogroups, M, N, and R: M was the source of C and D, N was the source of A and X, and R was the source of B.

All mitochondrial haplogroups from throughout the world—both African and non-African—have accumulated distinct, more recent variants that allow scientists to clearly distinguish subtypes within each haplogroup. I'll use the variants present in my own mitochondrial DNA as an example. According to a DNA test, I carry haplogroup U5, one of the most widespread and ancient mitochondrial haplogroups in Europe and in people whose maternal ancestry traces to Europe. It arose by mutation approximately thirty-six thousand years ago and was carried by people who spread throughout Europe at that time. My particular subtype is most prevalent in Scandinavia, the British Isles, and the northern parts of the European subcontinent. It also was found in ancient DNA extracted from a Stone Age individual who lived approximately 8,700 years ago and whose remains were discovered in the Hohlenstein-Stadel cave near the city of Ulm in Germany.[11] My mitochondrial lineage, when traced back to Africa, is the following: U5 (Europe) < U (Northern Middle East) < R (Middle East) < N (Middle East) < L3 (Africa).

The sort of hierarchical clustering evident in mitochondrial haplogroups allows scientists to extrapolate back to the trunk of the human mitochondrial family tree. Through comparison of mitochondrial DNA sequences from thousands of living people and the remains of ancient people and Neanderthals, scientists have reconstructed the original ancestral mitochondrial DNA sequence for all humans. This original mitochondrial DNA is no longer present in anyone alive today because variants have accumulated in everyone's ancestry, but everyone alive now traces her or his mitochondrial ancestry to a single woman who carried it long ago.[12]

Because each variant arose at one time in one person, any variant that is present in all people today but is not present in other closely related species, such as Neanderthals, chimpanzees, and gorillas, typically traces to one person. (There are rare exceptions, when the same variant arose on different occasions in different individuals, and these repeated variants can be identified by their presence against different genetic backgrounds.) For this reason, all ancestral lines of the human mitochondrial family ultimately lead to this one woman who lived in Africa nearly two hundred thousand years ago, according to a recent estimate, close to the time when anatomically modern humans first appeared.[13] She is famously known as the *mitochondrial Eve*, and there is no doubt she was African. The oldest variant in mitochondrial DNA that is inherited by all humans alive today, and by no other species, arose in her and was inherited by at least one of her daughters. She was not the first human female, however. Her mother, her mother's mother, her mother's mother's mother, and so on were all ancient humans and all mitochondrial Eves—ancient mothers of all humanity.

The Y chromosome in males is very different genetically from mitochondrial DNA, but its pattern of inheritance mirrors that of mitochondrial DNA: it is inherited through purely *paternal* lineages, exclusively from father to son. Like mitochondrial DNA, it does not recombine, so the same sort of hierarchical grouping of mitochondrial variants can also be done with Y chromosome variants.[14] And the same general patterns of diversity are present in DNA from the Y chromosome. Not surprisingly, the greatest diversity is in Africa, and the same general patterns of emigration within Africa and throughout the rest of the world are also apparent in Y chromosome DNA.

It is also possible to extrapolate back to the ancestral trunk of the Y chromosome family tree. All human males trace the DNA in their Y chromosome to a single man who lived in Africa probably a few tens of thousands of years later than the mitochondrial Eve—possibly about 142,000 years ago, according to a recent estimate, although dating methods for Y chromosome DNA are less reliable than for mitochondrial DNA.[15] He is known as the *Y chromosome Adam*. Obviously, given the time period in which he lived, he never met the mitochondrial Eve. In fact, there is a good chance he was one of her distant descendants, having inherited his mitochondrial DNA from her hundreds of generations later.

Although the most extensive studies of human diversity to date are from mitochondrial and Y chromosome DNA, scientists in recent years have extensively characterized human genetic diversity throughout all our DNA. The vast majority of our DNA is inherited from both parents and does recombine, so superimposition of recent variants on a background of ancient variants does not persist indefinitely as it does in mitochondrial and Y chromosome DNA. Nonetheless, some layering of variants that are near one another in each DNA molecule does persist for many generations, so each new variant, regardless of where it resides, is superimposed on a genetic background. And this variation throughout our DNA adds a mountain of evidence to the mitochondrial and Y chromosome evidence that our ancient origins are African.

Recently, tools have become available that allow scientists to examine DNA sequences on a massive scale in thousands of people, thanks to recent advances in large-scale DNA sequencing and computer technology. These large-scale studies have shown that humans carry an extraordinary number of extremely rare variants. Many of these rare variants are found in only one person of the thousands included in each study. These rare variants have recent origins, and the reason there are so many of them is simple and predictable: the worldwide human population has exploded in recent times, so the number of opportunities for new mutations to appear and become inherited variants has been greater in recent times than at any other time in human history.[16] Although they reveal much about recent human genetics, these extremely rare variants contribute very little to overall human diversity. It is the more ancient variants, distributed more widely throughout human populations, that account for the bulk of genetic diversity among humans and reveal continental-scale ancestry.

Jorde and Wooding, whose research we reviewed in the previous chapter, sum it up nicely. Referring to the DNA variants they examined, they state,

> All of these findings, which are in accord with many other studies based on different types of genetic variation assessed in different samples of humans, support an evolutionary scenario in which anatomically modern humans evolved first in Africa, accumulating genetic diversity. A small subset of the African population then left the continent, probably experienced a population bottleneck and founded anatomically modern human populations in the rest of the world. Of special importance to discussions of race, our species has a recent, common origin.[17]

According to the scientific evidence, the biblical Eve and Adam mentioned at the beginning of this chapter as the first parents of all humanity are mythical. Ancient human remains as well as DNA evidence show that humans were widespread throughout the world long before six thousand years ago, when, according to a literal interpretation of biblical history, Adam and Eve supposedly lived. The mitochondrial Eve and the Y chromosome Adam, however, were real, and they were African. Furthermore, they were not the only "Eves" and "Adams"—common ancestors of all humanity. All uniquely human variants in DNA present in all people alive today trace their origins to countless common ancestors, all of whom lived in Africa more than sixty thousand years ago. As humans, everyone is related by common ancient ancestry, and, ultimately, everyone is African.

CHAPTER 3

ANCESTRY VERSUS RACE

The Founding Fathers of the United States are widely revered for their intellect, courage, and foresight, which allowed them to lay the foundations of modern democracy. They lived during the Enlightenment—the age of reason—a period of intellectual thought based on the ideals of rationality, equality, and human rights that inspired the framework of democratic government and the philosophies that underlie modern scientific methods and inference.

Thomas Jefferson was among the most influential of the Founders. As author of the Declaration of Independence, congressman, minister to France, governor, secretary of state, vice president, and two-term third president of the United States, his mark on American history and modern political thought is unquestioned. That Jefferson was a champion of freedom and equality while being a slaveholder is one of the great enigmas of his life. After his wife, Martha, died, he inherited her slaves, one of them a woman named Sarah Hemings, who went by the name Sally. She was legally a slave, even though three of her four grandparents were of European ancestry; her maternal grandmother was an African slave, whereas her father and maternal grandfather were both European American slave owners. Interestingly, her father was also Martha's father, so Sally and Martha were half sisters even though Martha owned Sally as a slave. One of Sally's sons, Eston, took on the name Jefferson, and his descendants have carried the name for generations. A quiet tradition passed on for generations among Sally Hemings's descendants held that after Martha Jefferson's death, Thomas Jefferson cohabited with Sally and fathered all six of her children.

Powerful confirmation of this tradition came to light more than two centuries after Sally's death. The Jefferson Y chromosome haplogroup is a very

45

rare one, called T, which is found most often in Egyptian men but also in a very small fraction of European men. The European subtype is in Thomas Jefferson's paternal ancestry, and its prevalence is very rare among men of European ancestry. Eston Hemings Jefferson was Sally Hemings's last child, and he is the only one of Sally's children whose documented paternal lineage remains to the present. A paternal descendant of Eston Hemings Jefferson carries the same Y chromosome as Thomas Jefferson.

Historical information regarding Thomas Jefferson and Sally Hemings, particularly coincidence of the dates when Thomas was in the same place as Sally and the times when Sally's children were conceived, offers historical evidence that he was the most probable father of all six of Sally's children. The DNA evidence is consistent with this historical evidence.[1]

Sally Hemings's descendants are illustrative of how problematic definitions of race can be. All six of her children had substantial European ancestry but were still legally classified as slaves. Of the four who lived to adulthood, two left Thomas Jefferson's Monticello home as "fugitives" while in their twenties. The word "fugitives" is in quotation marks because there is good evidence that they were not only allowed but encouraged to leave and live in freedom, likely by Thomas Jefferson himself. Both changed their names and lived prosperously as members of white society. Eston eventually moved to Madison, Wisconsin, taking on the surname Jefferson, and his descendants were considered white, largely because of where they lived, how they were raised, and how they appeared. By contrast, some of his brother Madison's descendants were considered as colored and others as white. For instance, two of Madison's sons, Thomas and William, enlisted as Union soldiers in the US Civil War; Thomas was designated as colored and William as white.

Classification of people into discrete racial categories has a long history dominated by the dogma of white supremacy, based on interpretations of biblical history that were popular at the time. For example, a belief common to Christianity, Judaism, and Islam during medieval times was the notion that Africans are the descendants of Noah's son Ham and dark skin is the "curse of Ham" or "Hamitic curse."[2] Adherents of this belief divided the known world into three regions—Africa, Asia, and Europe—and assigned the supposed three major races to these regions as the posterity of the three sons of

Noah: Africans to Ham, Europeans to Japheth, and Asians to Shem (figure 3.1). Dark skin was thought to be God's curse on Ham for looking upon Noah when he was drunk and naked, as recounted in Genesis 9:18–25. This quasi-biblical scheme with its three-race classification was often used as justification for subjecting those who had supposedly inherited the curse of Ham to slavery.

Figure 3.1. *Mappa mundi* (map of the world) from *Etymologiae* by Isidorus, printed in 1472. The known world was thought to be surrounded by the ocean, with three continents populated by the descendants of Noah's three sons: Asia, populated by the descendants of Shem; Europe, populated by the descendants of Japheth; and Africa, populated by the descendants of Ham. Dark skin of Africans was supposedly the curse imposed on the descendants of Ham. *Image from the Harry Ransom Center, University of Texas at Austin.*

Also common in Europe and the Americas from the seventeenth century well into the twentieth century was the conjecture that the biblical mark of Cain is dark skin and that people with African ancestry are the seed of Cain. Connecting the mark of Cain to the curse of Ham was the presumption that Ham married a descendant of Cain, thus propagating the curse of Cain in his descendants after the great flood.[3] Adherents of these conjectures supported slavery based on the claim that God had cursed people with African ancestry and, by inheritance of this curse, they were inferior.

In the eighteenth century, scientists and scholars attempted to classify humans into distinct races, although on purely geographic rather than biblical grounds. The founder of modern biological taxonomy, Carl von Linné (better known by his Latinized name, Carolus Linnaeus), proposed in his 1758 work *Systema naturae* a Latinized classification of four human races: *Americanus*, *Europeus*, *Asiaticus*, and *Afer*—essentially the same as the three medieval classifications but with Native Americans added as a fourth category (the Americas were unknown to Europeans during medieval times). His follower, Johann Friedrich Blumenbach, expanded this classification into five so-called human varieties: Caucasian, Mongolian, Ethiopian, American, and Malaysian.[4]

Racial, ethnic, and cultural classification have persisted to the present. For example, the 2010 US Census listed five "racial categories" and allowed people to classify themselves into one or more than one category: 1) White, 2) Black or African American, 3) American Indian or Alaska Native, 4) Asian, and 5) Native Hawaiian or Other Pacific Islander. Regardless of which racial category a person chose, the census also allowed each individual to self-classify as Hispanic or non-Hispanic. It also included the following caveat: "The racial categories included in the census questionnaire generally reflect a social definition of race recognized in this country and not an attempt to define race biologically, anthropologically, or genetically."[5] This caveat is consistent with current scientific evidence suggesting that discrete racial classification is predominantly a social construct and does not reflect an accurate biological classification.

Complex ancestries often complicate attempts to biologically classify people by race. North Africa, the Middle East, and central and south Asia were anciently the crossroads of major civilizations, with people from various regions of the world traveling through, emigrating, or entering as traveling merchants or invading armies for thousands of years. As a consequence, genetic diversity for people with ancestry from this broad region is second only to sub-Saharan Africa. And there is so much overlap in the genetic variants carried by people from this region that genetic classification fails to place people into discrete racial categories.

In the postcolonial Americas, most people trace at least some of their ancestry to immigrant populations, often with varied ancestries. In Brazil, where immigrants from many regions of the world have converged and cultural taboos against so-called interracial marriage were not as powerful as in other cultures, ancestry is highly diverse. People in the United States who today classify themselves as African American often have some European ancestry—in many cases, significant European ancestry. For example, about 26 percent of African American men carry a Y chromosome of European origin.[6] Most people who classify themselves as Hispanic have European and Native American ancestry, and a large proportion have African ancestry. The same is true for many who self-identify as Native American. In fact, membership in Native American tribes in the United States is often disputed because of varied ancestry. Likewise, some who classify themselves as white or Caucasian (usually a catchall term to include people of European, Middle Eastern, and north African ancestry) have some African, Native American, or Asian ancestry. It is not unusual for someone who self-classifies as "white," "European American," or "Caucasian" to discover from a DNA test that his or her ancestry is varied. Had DNA tests been available and widely applied during the time of the antimiscegenation laws and the "one-drop rule" in the United States, a significant number of marriages considered valid at the time would have been technically illegal.

According to current scientific evidence, everyone's ancestry is ultimately African, and much of the variation in modern human DNA is original African variation dispersed throughout the entire human species rather than being confined to any particular geographic group. An example comes from research my colleagues and I recently conducted and published on the evolution of a human gene called *NANOG*.[7]

Biologists who discover a new gene have the honor of naming it, and, in this case, the honor fell to Dr. Ian Chambers of the University of Edinburgh in Scotland, who discovered the gene in mice in 2003.[8] He picked one of the best names in the history of gene naming, one that is both appropriate and charming. The name comes from *Tir na nÓg*, a mythical island in the sea west of Ireland that in Celtic legend is the land of eternal youth (also a popular name for Irish pubs). The gene functions in embryonic stem cells, which arise from cell division shortly after egg and sperm unite. When the *NANOG* gene is turned on, embryonic stem cells continue to grow and divide to make more embryonic stem cells rather than differentiating into different cell types. In other words, they remain indefinitely in a state of eternal youth, hence the mythical name for the gene.

Our research, however, was not on *NANOG*'s ability to maintain the eternal youth of embryonic stem cells; instead, we focused on the gene's evolutionary history in humans. While conducting our research, we found several ancient variants in this gene that, according to the evidence we discovered, diverged at least two million years ago, before the emergence of anatomically modern humans and at a time when all ancestors of modern humans lived exclusively in Africa. These variants are now spread among people throughout the world—in people native to Africa, Asia, the Middle East, the Pacific Islands, Europe, and the Americas, according to our research. This worldwide diversity persists because it was originally present in Africa.

Perhaps the best documented example of ancient African variation that is spread throughout the world is variation that confers the ABO blood groups: types A, B, AB, and O. There is a good chance you know your ABO blood type. For the purposes of this discussion, we'll ignore the positive and negative types, which arise from a different set of variants in a different gene. The ABO blood types arise from three major variants of one gene; the three variants are named A, B, and O. Divergence of the A and B variants is very ancient, having occurred at least twenty million years ago in a common ancestor of humans, great apes, and Old World monkeys.[9] This variation has persisted to the present, not only in humans but in other primates as well. Therefore, both the A and B variants are considered ancestral in humans. The O variant that causes type O blood in humans, however, is human specific. It is a derived variant that arose in one person in ancient Africa before the out-of-Africa diasporas, when the A variant mutated into the O variant. The mutation event deleted a single base pair (boxed) from the A variant to create the O variant:

A variant
TCCTCGTGGT**G**ACCCCTTGGC
AGGAGCACCA**C**TGGGGAACCG

O variant
TCCTCGTGGTACCCCTTGGC
AGGAGCACCATGGGGAACCG
deleted

This deletion completely obliterated any function for the O variant; it could no longer encode production of the substance that causes type A blood. The letter O is from the German word *ohne*, which means "without." More than four billion people throughout the world have type O blood, so, obviously, the lack of function conferred by the O variant is not detrimental for survival. In fact, as we'll see momentarily, there is some evidence that it anciently conferred an *advantage* for survival in some parts of the world.

Although all four blood types—A, B, AB, and O—are distributed among people worldwide, the relative proportions of these types differ geographically. Type O blood is the most common type worldwide, present in about 63 percent of the world's people, even though the O variant is derived from an ancient African mutation. Its predominance is particularly high in people whose ancestry is African, northeast Asian, or Native American, especially in Central and South America, where it approaches 100 percent among indigenous people. Type A is most common in people whose ancestry is Aboriginal Australian, northern European, or from the northern regions of North America. Type B is most frequent in people whose ancestry is from central Asia ranging from northern Russia to the southern tip of India. Type AB is the most rare type everywhere, and it happens only when a person inherits the A variant from one parent and the B variant from the other.

A complex pattern of ancient human emigrations, natural selection, and random fluctuations in the proportions of different variants accounts for the uneven distribution of ABO blood types. Natural selection has apparently been underway for all of human history, with different ABO blood variants conferring resistance or susceptibility to several diseases—in most cases, favoring type O blood and thereby explaining its high worldwide prevalence. The most important is protection against severe malaria. People with type O blood may contract malaria, but, once they have it, they are less likely to develop symptoms as severe as those who have other blood types. This observation may explain why type O blood is the most common type in Africa and other tropical regions, where malaria may have acted as a selective agent favoring survival of those people with type O blood who contracted malaria both before and since the out-of-Africa diasporas.[10] By contrast, people with type O blood may be more susceptible to cholera than those with type A, B, or AB, which explains the higher prevalence of the A and B variants in regions where cholera was historically common but malaria was not.[11]

The uneven yet worldwide distribution of A, B, and O variants is typical of ancient African variation in humans. The variants originated anciently in Africa and have spread throughout the world through complex human migrations. A combination of factors, including migration, natural selection, and random fluctuations, resulted in different proportions of the variants among people in different geographic regions.

Variants in DNA can generally be classified as three different types as a result of ancient human migrations: 1) ancient African variants present in people throughout the world, such as the ABO blood variants and *NANOG* variants we just discussed; 2) ancient African variants that remained predominantly or exclusively in people who were native to Africa until recent times; and 3) more recent variants that trace to a particular region, found in people with ancestry from that region. A subclass of this third class is very recent variants that are extremely rare and highly localized, and they are exceptionally numerous because of the rapid expansion of the world's population in modern times.[12]

Although most of the worldwide variants are the legacy of ancient African diversity, the recent localized variants tell us much about how recent diversity originated and what it means. The scientific literature is replete with examples, and my colleagues and I have discovered some of them firsthand in research we conducted with DNA from geographically diverse human populations.[13]

Another very important point is that most recent variants are not entirely confined to a limited geographic region. Instead, scientists usually look at *prevalence*: what proportion of people in any particular region carry a particular variant. The pattern that typically emerges is called a *clinal pattern*, in which the prevalence of a variant is highest in one particular region, then declines with increasing distance from that region, as illustrated in figure 3.2.

Figure 3.2. Clinal pattern for the prevalence of a variant in the gene *SLC45A2*, which influences skin pigmentation. This variant is most prevalent in Europe and declines gradually moving south and east.

The principal reason for clinal patterns is the combined effect of mutation and migration, sometimes coupled with natural selection. A variant originally arises by mutation in one person and is transmitted to at least some of that person's descendants. Over many generations, large numbers of that person's distant descendants may inherit the variant. And if it is favored through natural selection, its prevalence increases over multiple generations. Natural selection is not required, however; a variant's prevalence may also increase from one generation to the next purely by random fluctuations. If people who inherit that variant emigrate away from the region where their original ancestor lived, they carry the variant with them, and it becomes established elsewhere. In terms of probability, however, many variants (though not all) are most prevalent in the region of their origin.

Variants that appeared recently in human history tend to be localized geographically and often display clinal patterns. Although they are found in relatively small proportions of people, there are millions of these types of variants, and, thus, they offer plenty of opportunities for DNA analysis. When subjected to collective statistical analysis, they can identify geographic ancestries with high levels of certainty, revealing some fascinating information about our genetic histories. Some of these variants are so closely tied to ancestral geography that they are called *ancestry informative markers*. Each individual ancestry informative marker typically provides only meager evidence on its own about a person's ancestry. As one group of scientists who examined variants sampled from 383 indigenous people from various places in the world put it: "We found not a single [variant], out of nearly 250,000, at which a fixed difference would distinguish any pair of continental populations."[14] Rather, the cumulative information from thousands of variants provides reliable statistical probabilities for ancestry, be it from a single region or diverse lines of ancestry from different parts of the world.

For example, I've had my DNA tested for thousands of ancestry informative markers, and my recent ancestry (meaning the past ten thousand or so years) is almost entirely European. This conclusion, based purely on analysis of DNA variants, is fully consistent with my recorded genealogy, which is replete with British and Dutch surnames. And it has helped address some questions about my ancestry. For example, there is a long-held tradition in my extended paternal family that the surname Fairbanks was anglicized in fifteenth-century England in the region of York from the French surname Beaumont (*beau* means "beautiful" or "fair," and *mont* means "hillside" or "bank"). The People who immigrated to England with the Norman Conquest during the eleventh century presumably carried the name Beaumont, and their descendants later Anglicized it to Fairbanks (actually Fayerbanke at the time).

Potentially, I might confirm or refute this proposed history with Y chromosome DNA analysis. My Y chromosome haplogroup is one of the most common and widespread Y haplogroups in Europe. The subtype I have is most prevalent in northern France and Germany, the Netherlands, Belgium, Denmark, and southern England, concentrated near the shores of the North Sea. There is little doubt that men who were part of the Norman Conquest from France into England carried it, although it might have already been present in England prior to that time. Thus, the evidence is fully consistent with the tradition for how my surname originated, but does not prove it.

Although they constitute a minority of the millions of variants that define overall genetic diversity among humans, ancestry informative markers are useful for revealing a person's ancestry, for forensic purposes, and for identifying genetic variants associated with human health.[15] Variants associated with geography, such as those that cause variation in skin, hair, and eye pigmentation, are scientifically, socially, and politically important, in part because they dispel historic notions of racial superiority and help us to better understand our origins and diversity. Moreover, an understanding of how variants influence genetic conditions and disease, in the context of geography and human genetic diversity, are becoming increasingly useful for the prevention and treatment of disease through modern medicine.

The uneven distribution of genetic diversity throughout the world has important implications for the meaning of "race" in humans. The notion of three major races—African, European, and Asian—makes little biological sense because those three groupings are far from equal in terms of diversity, and there are no distinct genetic lines that separate them. If genetic diversity were the sole basis of racial classification, different geographic and ethnic groups in sub-Saharan Africa should be divided into a much larger number of races than geographic and ethnic groups throughout the rest of the world, due to the higher degree of ancient genetic diversity in Africa.

Doing so, however, would also make little biological sense because so-called ethnic groups represent complex two-dimensional continua of genetic overlap for literally millions of variants. When geneticists sample DNA from indigenous people living in discontinuous geographic extremes, such as northern Europe, central Africa, and east Asia, they can use ancestry informative markers to readily define these groups. However, this simple definition is an artifact of discontinuous sampling, not a true representation of the world's genetic diversity. When sampled over a wider range of geography, DNA variants that are common in people from a certain region may be found in lower proportions of prevalence elsewhere, with much overlap for different variants. Human migrations have distributed and mixed the world's genetic variation, and mutations have added to that variation throughout the course of human history. Geographic ancestry for any particular person is best defined statistically by the combination of numerous ancestry informative markers each person carries.

What this means scientifically is that classification by race is an oversimplified and inaccurate way to biologically define people. Instead, some people have lines of ancestry that may be highly diverse, tracing to different parts of the world, whereas others have more narrow ancestry, localized to a few nearby regions or, more rarely, to a single one. For instance, my DNA-based ancestry is narrowly European, whereas a friend's DNA analysis revealed a combination of significant proportions of European, Native American, north African, and west Asian ancestry. Both of us would be classified as "white" and "not Hispanic" under the US Census classification scheme. Moreover, because current classification schemes are based entirely on self-reporting, each person's classification is determined by her or his self-perception of race rather than by any sort of scientific analysis.

Ancestry, rather than race, is what defines each of us biologically. And a statistical analysis of the combination of DNA variants each person carries makes it possible to identify with a high degree of confidence the biogeographical origins of her or his lines of ancestry. Even so, social or cultural ancestry often means much more to people than biogeographical origins, and it may not necessarily be the same as biological ancestry. As stated by the Race, Ethnicity, and Genetics Working Group of the National Human Genome Institute:

> At least among those individuals who participate in biomedical research, genetic estimates of biogeographical ancestry generally agree with self-assessed ancestry, but in an unknown percentage of cases they do not.

> Despite its seemingly objective nature, ancestry also has limitations as a way of categorizing people. When asked about the ancestry of their parents and grandparents, many people cannot provide accurate answers. . . . Misattributed paternity or adoption can separate biogeographical ancestry from socially defined ancestry. Furthermore, the exponentially increasing number of our ancestors makes ancestry a quantitative rather than a qualitative trait—five centuries (or twenty generations) ago, each person had a maximum of >1 million ancestors. To complicate matters further, recent analyses suggest that everyone living today has exactly the same set of genealogical ancestors who lived as recently as a few thousand years in the past [about two hundred thousand years ago in Africa], although we have received our genetic inheritance in different proportions from those ancestors.

In the end, the terms "race," "ethnicity," and "ancestry" all describe just a small part of the complex web of biological and social connections that link individuals and groups to each other.[16]

The time has come to abandon the notion of race as a presumed biological construct when referring to humans. It may be legitimately argued that terms such as *race* and *ethnicity* have value as social constructs. But when referring to a person's genetic constitution, we should turn our attention to *ancestry*, which is scientifically more informative, less burdened by political and historical baggage, and immensely more complex and fascinating.

CHAPTER 4

"THE COLOR OF THEIR SKIN"

A few years ago, I visited the Lincoln Memorial in Washington, DC, early on a cold January morning. Except for a guard sitting in his warm booth in a dark corner of the building, I was alone. I gazed at Daniel Chester French's monument of Abraham Lincoln carved in Georgia marble, admiring the rugged lines French had so skillfully captured in Lincoln's face. I then turned around, looking east toward the Washington Monument. At my feet, I noticed a small brass plaque embedded in the marble step, identifying that place as the site where Martin Luther King Jr. delivered his most stirring speech: "I Have a Dream." His familiar words entered my mind, along with my childhood recollection of the penetrating sound of his voice:

> Now is the time to rise from the dark and desolate valley of segregation to the sunlit path of racial justice. . . . I have a dream that my four children will one day live in a nation where they will not be judged by the color of their skin but by the content of their character.[1]

I chose his words, "the color of their skin," as the title for this chapter because no other human characteristic is so strongly associated with the perception of race. The names of colors—black, white, red, and yellow, among others—have been used as legal definitions of race, as labels of supposed superiority or inferiority, and as common descriptions of people whose ancestry derives from various regions of the world. Since I was a child, I've always been puzzled by these color characterizations of people, in light of the obvious fact that actual variation for skin color in humans does not fall into discrete classes, nor is it actually white, red, yellow, or black. Instead, it ranges from intense to little pigmentation in continuously varying gradations.

The genetic basis for variation in human skin pigmentation, as well as hair and eye pigmentation, is now quite well understood. Current scientific evidence paints a clear picture of how, when, and where DNA variants arose and why patterns of genetic variation that confer pigmentation are distributed as they are throughout the people of the world. That variation for pigmentation is largely inherited is beyond question, a fact well known for centuries, long before the scientific principles of inheritance in humans were understood. There is, of course, some influence from environment, such as the protective response of increased pigmentation in some people when their skin is repeatedly exposed to sunlight, popularly known as tanning. Yet even the ability to tan is itself an inherited characteristic. For the most part, variation for skin, eye, and hair pigmentation is a consequence of genetic ancestry.

Inheritance of skin color is complex, conferred by variants in a relatively large number of genes, a situation scientists call *polygenic inheritance*. The polygenic nature of variation for skin color in humans has been known for some time; I recall studying the evidence of it in the first college biology course I took, in 1978. However, until recently, most of the individual genes, and the specific variants in their DNA that influence variation for skin pigmentation, remained unknown. Now, many of those genes and their variants have been identified, and scientists have examined them in detail, discovering how they regulate pigmentation and how they are distributed geographically. Before examining these variants individually, let's review some of the major conclusions from this research.

First, high pigmentation in the skin, hair, and eyes is the ancestral state of all humans. Current evidence from DNA overwhelmingly confirms this conclusion, and it fits a common pattern in genetics. Most mutations in DNA that have any effect on a gene reduce or eliminate the function of that gene. The reduction of pigmentation in people whose postdiaspora origins lie outside of Africa is due to ancient mutations in these genes. The variants that arose from those mutations cause a reduction in gene function, which reduces pigmentation, resulting in the variation currently present in modern humans.

The second conclusion of this genetic research is that most of the mutations that became variants affecting skin, eye, and hair pigmentation happened outside Africa in the distant descendants of people who originally left Africa. Unlike variation for blood groups (such as A, B, AB, and O blood), which vary among people throughout the world, most variants that influence pigmentation are not original African variation. Instead, the specific variants responsible for reduction in skin, hair, and eye pigmentation arose as mutations in individual people and then spread through their descendants within broad geographic regions. These variants constitute some of the most reliable ancestry informative DNA markers.

The third conclusion is that by examining the genetic background of any particular variant, scientists can detect patterns in DNA that indicate that Darwinian natural selection has influenced the prevalence and geographic distribution of each variant. Of the many contributions Darwin made to science, the one that most impacted human thought is the principle of *natural selection*. His definition of it in the *Origin of Species* is perhaps the best ever penned:

> Any variation, however slight and from whatever cause proceeding, if it be in any degree profitable to an individual of any species . . . will tend to the preservation of that individual, and will generally be inherited by its offspring. The offspring, also, will thus have a better chance of surviving, for, of the many individuals of any species which are periodically born, but a small number can survive. I have called this principle, by which each slight variation, if useful, is preserved, by the term of Natural Selection.[2]

There is clear evidence that variants in DNA conferring variation for skin pigmentation in humans, ranging from very dark to very light skin, have been subjected to intense natural selection. In some cases, the effect of selection was relatively rapid, taking place over a few hundred generations instead of a few thousand. Such rapid effects of natural selection are called *selective sweeps*. For example, several variants in humans conferring resistance to disease were subject to rapid selective sweeps. People resistant to the diseases survived and reproduced, whereas those susceptible to these diseases tended to succumb and die before reproducing. The prevalence of any variant conferring resistance to disease, therefore, rapidly increased from one generation to the next through natural selection. Recognizable patterns in human DNA provide evidence that some variants known to reduce skin pigmentation have also been subjected to rapid selective sweeps, resulting in an increase in their prevalence. In fact, some of the most powerful selective sweeps in ancient human history appear to be associated with variation for skin pigmentation.

The fourth conclusion is that the geographic distribution of skin pigmentation in humans is best explained through a pattern of natural selection superimposed on ancient human migrations, called the *vitamin D hypothesis*. When ancestral human populations are examined, pigmentation is greatest in equatorial Africa and tends to decrease with increased distance from the equator. The most likely environmental agent responsible for this pattern is the amount of winter sunlight to which ancient people were exposed. Pigmentation protects the skin from ultraviolet radiation when sunlight is intense; this is the principal biological function of skin pigments. In regions where sunlight is intense throughout the year, high pigmentation confers a selective advantage to people who have it because it protects against ultraviolet light-induced degradation of folate, a substance that is essential for fetal development. This is especially important in pregnant women because folate protects the fetus from disabling birth defects.[3]

But, according to the vitamin D hypothesis, there is a natural tradeoff. The skin is the place in our bodies where vitamin D is produced, and exposure to sunlight stimulates vitamin D production. High pigmentation in the skin can inhibit vitamin D production when sun exposure is reduced. In regions where sunlight is intense throughout the year—regions near the equator—exposure to sunlight was sufficient to ensure more than adequate vitamin D production in ancient people, while the inhibitory effect of skin pigmentation also protected against folate degradation. However, when the descendants of people who left Africa immigrated into the more northern latitudes of Europe and east Asia, the world was in the midst of a major ice age, which reached maximum cooling and glaciation at about twenty-two thousand to twenty-four thousand years ago, about the time when people were colonizing these regions. Winters were colder than they are now, so people in these regions had to cover most of their bodies and seek shelter to protect themselves from the cold, depriving them of already meager winter sunlight. Even more then than now, short winter days in northern latitudes and overcast skies limited the amount of sunlight during the winter months. Insufficient exposure to sunlight resulted in vitamin D deficiency in these people, causing a disease known as rickets, because those with highly pigmented skin could not produce enough vitamin D during the winter season. The most serious symptom of rickets is weak and easily fractured bones. Pregnant women were especially affected, often culminating in death for them and their unborn children. People with lower amounts of skin pigmentation, however, had a strong selective advantage because their bodies were better able to produce vitamin D with limited sunlight than people with higher skin pigmentation. And the lack of ultraviolet light exposure also protected against folate degradation. Rapid selective sweeps caused variants that reduced skin pigmentation to spread over a period of generations in people living in northern regions— in some cases, entirely displacing original ancestral variants in people who lived in the northernmost regions of Europe. And the farther north people migrated, the more variants conferring reduced pigmentation accumulated over generations through a combination of mutation and natural selection.

There are explainable exceptions to this general rule. For example, Native Americans in coastal regions of Alaska, northern Canada, Greenland, and northeastern Siberia—collectively known as Eskimo people—have lived in high-latitude, low-sunlight environments for thousands of years. Yet light skin complexions have not evolved there. In this case, the most probable reason also has to do with vitamin D. Diets in these regions since ancient times have been high in seal liver and fish, which naturally contain high amounts of vitamin D, compensating for reduced vitamin D production in the skin. With adequate vitamin D in the diet, reduced pigmentation did not confer a strong selective advantage.[4]

Although rare, cases of rickets are still reported when people fail to consume a diet with sufficient vitamin D and when sunlight exposure is insufficient for adequate vitamin D production in the skin. This is especially true for infants, children, and adolescents, whose bodies are growing and require proportionally more vitamin D than adult bodies. The American Academy of Pediatrics recommends vitamin D supplementation in the diet to prevent rickets.[5]

Skin damage resulting from overexposure to sunlight, however, remains a serious health issue. Sunlight-induced skin cancer does not explain why natural selection favored reduced skin pigmentation in northern regions (this type of cancer typically appears well after people have reproduced), but repeated exposure to sunlight can be a serious danger to individuals. Over time, it increases the probability of lethal skin cancer and other long-term skin disorders, especially for people whose skin complexions are light. Although skin pigmentation protects against sun damage, even high skin pigmentation is not an absolute protection. All people should protect themselves from overexposure to the sun. Nonetheless, light skin complexion is the highest genetic risk factor for malignant melanoma, the most serious and potentially lethal form of skin cancer.

Having reviewed the main conclusions of research on how and why variation for pigmentation in humans originated, let's dig a bit deeper, examining some of the direct evidence of how this variation, and its geographic distribution, evolved throughout human history. To understand this variation, we need first to understand what skin pigmentation is.

Human skin, hair, and eyes contain a group of related pigments called *melanins*, the prefix *melan-* meaning "dark" or "black." The word *melancholy*, meaning a dark and somber mood, has the same root. Melanins ultimately arise from the food we eat—specifically, from the protein in our food. Each individual protein molecule is structured like a linked chain that, when stretched out, becomes linear. The links that make up each protein chain are called *amino acids*. When you eat anything with protein (which is in almost everything you eat), your digestive system breaks down each protein chain into the individual amino acids, which your digestive system then absorbs into the bloodstream. Your cells extract the amino acids from the blood and reassemble them into new chains in new combinations to form your own proteins.

For instance, imagine eating a bowl of rice. The rice consists of seeds that contain mostly starch but also protein and a small amount of fat and fiber, along with trace amounts of vitamins and minerals. The natural function of proteins in rice is to provide a source of amino acids for a rice plant germinating from the seed so that the developing seedling can use the amino acids from those seed proteins to make its own proteins during its earliest growth stages. But by eating the rice seeds, you have co-opted those seed proteins for yourself. Your digestive system breaks down the rice proteins into individual amino acids, most of which end up in your bloodstream to be used by your cells to make proteins, such as the proteins that constitute much of your hair, fingernails, muscles, and blood. The proverb "You are what you eat" is correct, but it's more accurate to say, "You are a recombination of what you eat."

Although your cells use those amino acids to make your proteins, they divert some of them toward other purposes, one of which is to make melanins in the skin, hair, and eyes. Two amino acids—phenylalanine and tyrosine—are the starting points and are converted into melanins through a series of steps called a *pathway*. In each step of the pathway, one substance is chemically transformed into another substance. Figure 4.1 is a simplified depiction of some of the major steps in the melanin-synthesis pathway.

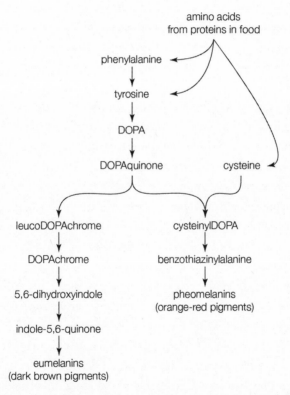

Figure 4.1. Major steps of the melanin biosynthesis pathway.

Genes in DNA govern each of the steps in the pathway, and those steps can only proceed when those genes are active and functioning. A mutation that causes a variant in any of these genes tends to reduce the activity of that gene, which, in turn, reduces the production of the melanins. The fact that these variants in DNA are inherited explains why the color of skin, hair, and eyes is inherited.

Let's look first at some obvious examples of what happens when particular variants are inherited, starting with the second and third steps of the pathway: the conversion of tyrosine to DOPA, and DOPA into DOPAquinone. Variants affecting these two steps have the most dramatic inherited effect possible on pigmentation. The gene that governs these steps is called *TYR*, and variants that completely disable this gene cause albinism—more accurately, oculocutaneous albinism—which means those who have this form of albinism have no pigment in the skin, hair, and eyes. The hair is white, the skin has an extremely light complexion, and the irises in the eyes appear red or pink because, in the absence of melanins, the red color of the blood vessels is visible in the iris.

Albinism is an uncommon inherited condition in humans, and it causes extremely high susceptibility to skin cancer and other related sunlight-exposure disorders, due to the absence of protective melanins in the skin. Mutations that cause albinism have occurred independently in humans at different times during human history and among various human cultures. Search the internet for images using the terms *albinism* and *human*, and you'll find dozens of photographs depicting people with albinism from various parts of the world.

Albinism is also well known in many species of animals, such as albino rabbits, mice, and rats raised as pets or laboratory animals. They, too, have white hair and red or pink eyes. Albino reptiles, birds, and fish are also well documented. Not surprisingly, variants in the precisely the same gene (*TYR*) in humans and other animals cause albinism in all of them.

After DOPAquinone, the pathway splits into two major branches that result in two major types of melanins: one called *eumelanins*, consisting of a group of dark bluish-brown pigments, and the other called *pheomelanins*, consisting of reddish-orange pigments. People who have blue eyes have reduced amounts of the bluish-brown eumelanins in the iris, hence the blue or bluish-green color. People who have what we call "red" hair carry one or more variants in their DNA that substantially reduce the bluish-brown eumelanins, so that the reddish-orange pheomelanins are the predominant pigment in the hair. (The same is true for reddish-orange coat color in orangutans, albeit by a different genetic process than in humans).[6] Notably, most variants that result in red hair reduce eumelanins in not only the hair but also the skin and eyes. Hence, people with red hair often have light skin complexions and either blue or green eyes, although not always.[7]

The wide range of pigmentation in humans is a consequence of variants in the numerous genes that regulate the pathway for melanin production. Each person carries a particular combination of variants in those genes, including ancestral variants that contribute to high pigmentation and, in many people, derived variants that reduce pigmentation. And the particular combination of variants a person carries determines the color of her or his skin, hair, and eyes. Because multiple genes regulate the various steps in the branching pathway for melanin synthesis, and because variations in any of these genes can influence melanin production, the inheritance of skin, hair, and eye color is highly varied and complex.

But not too complex for scientists to decipher, especially with the research tools now available. Variants that reduce pigmentation in people from different parts of the world are now well documented, and, undoubtedly, others remain to be discovered. Those variants tell an intriguing story, one that explains not only how variation for pigmentation arose but also why it is distributed throughout the world as it is.

Let's look at some examples. Among the many variants that influence pigmentation in humans, not all are equal. Some confer major effects on pigmentation, whereas the effect of others may be relatively minor. Also, because pigmentation reduction is greatest in people whose ancestry derives from northern latitudes, much of the research on the genetic basis for pigmentation reduction has focused on variants that originated in European and east Asian populations. We'll look at several of the well-researched major variants and the times and places where these variants originated.[8]

However, let's first turn to equatorial Africa, where the ancestral state of high skin pigmentation prevails. Is there any evidence in DNA that natural selection has preserved the highly pigmented skin of native Africans? Recall that sub-Saharan Africa is the region with the highest genetic variation in the world. Thus, we might expect that variation in the genes that govern skin pigmentation would be highest in people native to Africa. In fact, the opposite is true: the ancestral variants in genes governing skin pigmentation tend to be highly *uniform* in native Africans. As an example, the *MC1R* gene has a major effect on pigmentation, and variants in it are widespread throughout the world outside Africa, causing substantial reductions in pigmentation. For instance, one of these variants originated in what are now Scotland and the northern part of Ireland during an ice age when sea levels were lower and Ireland and Scotland were part of the same land mass. This variant results in very light skin complexion, freckling, and red hair and is most common in people whose ancestry traces to Ireland and Scotland.[9] By contrast, the ancestral variant for *MC1R* is almost uniformly present in native Africans, in spite of high variation in genes unrelated to pigmentation. In equatorial Africa, variants that reduced pigmentation were eliminated through natural selection, and the original ancestral variants that conferred dark pigmentation were preserved.[10] This type of natural selection—preservation of ancestral variants at the expense of derived variants—is called *purifying selection*, and it is common for many genes.

It is often said that the increased risk of skin cancer is the reason why natural selection has favored dark skin pigmentation in equatorial parts of the world. The scientific evidence suggests, however, that skin cancer is not the reason for natural selection favoring highly pigmented skin. Although potentially lethal and a serious risk for people with light skin complexions, skin cancer typically appears later in life after many years of sun exposure, usually well after people have reproduced. For natural selection to be effective, it must preserve variants that confer an advantage for survival *before reproduction*. Its effect on preserving variants that reduce survival *after reproduction* is substantially minimized. In the case of equatorial Africa, the selection agent instead appears to be degradation of folate by high exposure to ultraviolet radiation in sunlight. A process called *folate photolysis* happens when folate in the skin is exposed to ultraviolet radiation and the folate is degraded. Folate is essential for fetal development, and insufficient folate often results in birth defects and fetal mortality, clearly influencing reproduction. High skin pigmentation has protected people in equatorial regions against folate degradation throughout human history.[11]

Let's now turn our attention to variants that reduce pigmentation in people whose ancestry lies outside Africa. Recall that the out-of-Africa emigrations that founded non-African human populations took place about seventy thousand to sixty thousand years ago. The descendants of these original emigrants increased in number through numerous generations, and many of them began emigrating farther away from Africa. One group emigrated northward, eventually establishing settlements in an area near the Caucasus Mountains between the Black and Caspian Seas in what are now parts of the nations of Azerbaijan, Georgia, and Russia. This area became a major staging ground for subsequent immigrations into Asia and Europe, immigrations that founded the first human populations in these regions.

Interestingly, the term *Caucasian* is derived from this region of ancient origins. The term itself, however, is less than scientific. As quoted by Stephen J. Gould in his book *The Mismeasure of Man*, Johann Friedrich Blumenbach explained his invention of the term *Caucasian* thusly: "I have taken the name of this variety [of humans] from Mount Caucasus, both because its neighborhood, and especially its southern slope, produces the most beautiful race of men, and because . . . in that region, if anywhere, we ought with the greatest probability to place the autochthones [original forms] of mankind."[12] These views—that so-called Caucasians were the "most beautiful race" and that humans originated in that part of the world—were generally accepted among Europeans and European Americans at the time (late 1700s in the case of Blumenbach), as well as for centuries before and after.

Although Africa is undoubtedly the place where modern humans originated, the Caucasus region was a major center of secondary origin for the first European and east Asian populations. It is in this region that one of the oldest, most widespread, and most significant variants responsible for reduction in skin pigmentation appeared more than thirty thousand years ago. A mutation in a gene known as *KITLG* arose in one individual and was passed on as a variant to his or her offspring (there is no way to know whether the mutation happened in a female or a male). The variant that arose from this mutation reduced the skin color of those who inherited it, and it spread among the descendants of the individual in whom it first appeared.

The derived variant differs from the ancestral variant only by a single base-pair substitution: an A–T pair in the ancestral variant mutated into a G–C pair in the derived variant:

Ancestral variant
AAAAACTGAA**A**GATATTATTA
TTTTTGACTT**T**CTATAATAAT

Derived variant
AAAAACTGAA**G**GATATTATTA
TTTTTGACTT**C**CTATAATAAT

The origin of this derived variant before thirty thousand years ago predates the divergence of European and east Asian populations, and many people with ancestry from either of these regions now carry it. It also predates the climax of the most recent ice age, about twenty-four thousand to twenty-two thousand years ago. During this ice age, people who carried this variant had a distinct advantage for survival and reproduction as they migrated west into Europe or east into Asia. Because this variant was favored through natural selection in regions of lower sunlight, it is now most prevalent in people whose ancestry traces to Asia or Europe. In European populations, 84 percent of *KITLG* variants are this derived variant, and in east Asians, the percentage is similar at 82 percent. The ancestral nonmutant variant is still present in some people from these regions, but at low percentages. By contrast, the ancestral variant is by far the most common variant in indigenous African populations.

Another derived variant that substantially reduces pigmentation is in the gene *SLC24A5*. The derived variant arose from a mutation that changed a G–C pair to an A–T pair in the DNA of this gene:

Ancestral variant
TGTTGCAGGC**G**CAACTTTCAT
ACAACGTCCG**C**GTTGAAAGTA

Derived variant
TGTTGCAGGC**A**CAACTTTCAT
ACAACGTCCG**T**GTTGAAAGTA

The mutation that produced this variant arose between eleven thousand and nineteen thousand years ago in a person who lived somewhere between the Caucasus region and Europe and whose descendants immigrated into Europe, so this variant is most prevalent in people with European ancestry.[13] The DNA sequences on both sides of it bear variants that portray the marks of a powerful selective sweep that caused its rapid spread among the descendants of ancient European immigrants. As a result, nearly 100 percent of people whose ancestry is entirely European carry this variant. In fact, it is a variant considered to be one of the most reliable ancestral informative markers for European ancestry. It is exceptionally rare in all other native populations.

Because the derived variant is present in almost 100 percent of people whose ancestry is entirely or almost entirely European, it is not possible to tell how much it influences skin pigmentation by direct measurements in these populations; variation is essential to detect a correlation. However, most people whose ancestry is African American or Caribbean American have both African and European ancestry, so some carry two copies of this variant (inherited from both parents); some, one (inherited from one parent only); and others, none (not inherited from either parent). In these populations, the presence of this variant has a strong dosage effect on reduced skin pigmentation. Those who carry two copies of the ancestral variant have more pigmentation than those who carry one copy of the ancestral and one of the derived variant, and they have more pigmentation than those who carry two copies of the derived variant.

Two other derived variants that also reduce the amount of skin pigmentation show a similar pattern to this one in that they are almost uniformly present in European populations and uniformly absent in African and east Asian populations, and they arose in Europeans at about the same time. They, too, result from single base-pair substitutions in two genes: *SLC45A2* and *TYRP1*. Because the patterns are so similar, we'll not take the time to examine them in detail, other than to note that they, too, bear the hallmarks of selective sweeps that culminated in their high prevalence in ancient European populations.

A moment ago, we discussed the derived variant in the gene *KITLG*, which appeared early in what is called a proto-Eurasian population before the divergence of European and east Asian migrants. Thus, it reduces skin pigmentation in large proportions of people whose ancestry is either European or east Asian. Other, more recent, variants also reduce pigmentation in east Asian populations, and they are entirely independent of those in European populations.

Perhaps the best-studied example is a derived variant in the gene *OCA2*, which has one of the most significant and widespread pigment-reducing effects in people of east Asian ancestry. A mutation changed an A–T pair in the ancestral DNA to a G–C pair in the derived variant:

Ancestral variant

```
CTCTTACAGCATAGGATATCT
GAGAATGTCGTATCCTATAGA
```

Derived variant

```
CTCTTACAGCGTAGGATATCT
GAGAATGTCGCATCCTATAGA
```

This variant is common in east Asian populations, and, like other such variants, the DNA surrounding it bears the evidence of it being favored through natural selection.

The observation that natural selection has independently favored different pigment-reducing variants in European and east Asian populations is an excellent example of what biologists call *convergent evolution*. It results when the same external characteristic evolves independently in different populations when those populations are subjected to similar environmental conditions. The underlying variants responsible for that characteristic usually are different in those populations, as is the case here, because the mutations happen independently. In other words, the response to natural selection is the same in different populations, but the genetic basis is different.

Although this particular variant is east Asian in origin, some people of European ancestry also carry a different variant in this same gene (*OCA2*), and its origin is independent of the east Asian variant. Genes are long strings of base pairs in DNA, and a mutation can happen anywhere in a gene. Hundreds or even thousands of different variants in a single gene are possible when

the world's entire human population is considered. Other independent pigment-reducing variants are known in east Asian populations, including major
pigment-reducing variants in the genes *DCT* and *ATRN*, evidence that the
genetic basis for pigment reduction in east Asian populations is complex and
due to variants in several genes.

One of the most intriguing examples of convergent evolution for skin
pigmentation, however, comes from Neanderthals. In modern humans,
several derived variants in the *MC1R* gene are known to substantially reduce
eumelanins, resulting in red hair and very light skin pigmentation, often
with freckling. Variants in this gene are especially common in people with
Irish, English, and Dutch ancestry, due to mutations that happened anciently
in people living in northern Europe, and they cause red hair, as mentioned
earlier in this chapter.[14] In 2007, scientists from Spain, Italy, Germany, and
France discovered that DNA from the remains of a Neanderthal man, found
in the Italian Alps, contains a variant in this same gene that should have substantially reduced eumelanin production in this man.[15] According to their
analysis, this Neanderthal man probably had red hair and very low skin pigmentation, the same as in humans today who inherit similar derived variants
of this gene. Thus, natural selection may have favored low skin pigmentation
in at least some Neanderthals who lived in regions of Europe with low winter
sunlight. Although this variant is in the same gene as pigment-reducing variants in modern humans whose ancestry traces to Ireland and Scotland, this
particular variant has never been found in modern humans and is probably
exclusive to Neanderthals. It therefore must have arisen separately and independently, an example of convergent evolution for reduced skin pigmentation
in Neanderthals and modern humans. Interestingly, the ancestral variant for
this gene has since been discovered in DNA from other Neanderthals, suggesting they may have had greater pigmentation in their skin, hair, and eyes
than this individual—evidence that pigmentation varied in Neanderthals, as
it varies in modern humans.[16]

The ability to tan when exposed over a period of days to increased sunlight is also an inherited characteristic that has evolved through natural selection. The lightest complexions have evolved in extreme northern regions,
such as Scandinavia, northwestern Russia, the British Isles, and Ireland. In

these regions, winter sunlight is brief, and summer sunlight, though extended through long days, is not as intense as elsewhere in the world because of the sharper angle of the sun's rays relative to the surface of the earth. People with ancestry entirely or predominantly from these regions often sunburn easily and are unable to tan. Their extremely light skin complexions are an excellent adaptation to their region of ancestry because such complexions facilitate vitamin D production with low sun exposure. At more intermediate latitudes, a relatively light skin complexion during winter, when sunlight is weak, and a darker complexion during summer, when sunlight is intense, provide the combination of exposure and protection that is needed to facilitate the trade-off between vitamin D production and protection against folate degradation. The ability of people whose ancestry traces to intermediate latitudes to tan allows their skin to adapt to seasonal changes in sunlight. Some people have argued that the gradually increasing exposure to light required for tanning is not a particularly good adaptation because it requires so much time and initially does not prevent skin damage. In response, Nina Jablonsky and George Chaplin of the University of Pennsylvania wrote,

> Tanning is viewed by modern clinicians as an imperfect adaptation to UVR because it damages the skin's connective tissues, immune system, and DNA, and thus leads to progressive changes resulting in skin cancer. This is an appropriate statement for vagile and longevous 21st century humans but not for those of the 18th century or earlier who lived before the advent of widely available, rapid long-distance transportation. With early reproduction and before the extension of the average human lifespan through improvements in diet and medicine, skin cancer had no effect on reproductive success. Further, the genetic pattern of skin cancer risk does not accord with predictions based on selection for resistance to skin cancer. In the context of human evolution, the evolution of tanning was a superb evolutionary compromise.[17]

To conclude, in spite of the complexities of the genetic basis for skin color variation in humans, the inferences from scientific research are straightforward and unambiguous. Intense skin pigmentation is the ancestral state of humanity, and it traces to the original African origin of all humans. Variants conferring reduced skin pigmentation are strongly associated with ancient immigration during the most recent major ice age into Europe and east Asia, regions of low winter sunlight. People who inherited these variants had a survival and reproductive advantage in these regions because they were better able to produce vitamin D than people with higher skin pigmentation. By contrast, high skin pigmentation was advantageous to people who lived in regions of intense year-round sunlight nearer the equator because of the protective effect against folate degradation, which adversely affects fetal development. Many of the variants responsible for reduced skin pigmentation have been identified in people of European and east Asian ancestry, and they bear the molecular signatures of natural selection, including selective sweeps that rapidly increased the prevalence of these variants.[18]

Returning to Martin Luther King's words that opened this chapter, we have not yet reached the day when people are no longer "judged by the color of their skin." There is no evidence whatsoever from science to justify discrimination on the basis of skin color. On the contrary, science has given us a clear understanding of why variation for skin, hair, and eye pigmentation is such an obvious indicator of human diversity, and it has nothing to do with notions of superiority or so-called racial purity. Instead, it is clearly tied to our evolutionary history and the effects of natural selection, aspects of human diversity that should evoke a sense of wonder for the forces of nature that have shaped our past and present.

CHAPTER 5

HUMAN DIVERSITY
AND HEALTH

The afternoon of September 24, 2006, was hot and humid in Houston, Texas, not unusual for that time of year. It was the day after a Rice University football game, and those members of the team who had played in little or none of the game, mostly underclassmen, participated in a practice session with vigorous weight training and sixteen consecutive hundred-yard sprints. One of the players who practiced that afternoon was a nineteen-year-old defensive back who collapsed and lost consciousness after the final sprint. The following day, he died. Subsequent tests showed that he had sickle-cell trait, a typically benign genetic condition that manifests symptoms only rarely, if at all. When symptoms do appear, the trigger is usually a combination of extreme physical stress, heat exhaustion, and dehydration. In rare cases, the symptoms are severe and sometimes fatal, as in this tragic instance.

This young man was African American. Although sickle-cell trait occurs in people with many different ancestral backgrounds, it is most frequent in people with African ancestry. A derived variant in a gene called *HBB* is responsible for the condition:

Ancestral variant
```
CTGACTCCTGAGGAGAAGTCT
GACTGAGGACTCCTCTTCAGA
```

Derived variant
```
CTGACTCCTGTGGAGAAGTCT
GACTGAGGACACCTCTTCAGA
```

The reason for its relatively high prevalence in people with recent African ancestry is one of the most extensively researched and best-documented examples of natural selection in humans, with evidence dating back to 1954.[1] The ancestral variant is usually designated as *A* and the derived variant as *S*. Humans inherit two copies of each gene, one from each parent, so each person has one of three possible combinations of the *A* and *S* variants: *AA*, *SS*, or *AS*. In genetic terms, people who inherit the same variant from both parents (*SS* or *AA*) are said to be *homozygous*, whereas those who inherit different variants (*AS*) are *heterozygous*. People who are homozygous *SS* suffer from a serious genetic condition called *sickle-cell anemia*, characterized by severe fatigue due to a lack of functional red blood cells; intense pain, especially in the joints; swollen hands and feet; spleen damage, resulting in frequent infections; and damage to the retinas in the eye, resulting in impaired vision. Prior to modern medicine, it was often fatal during childhood or adolescence, but with current treatments, people who have it may live much-extended lifespans, although not without significant pain and suffering. Those who are heterozygous *AS* have *sickle-cell trait* (not sickle-cell anemia), and most never know they carry it because symptoms are usually nonexistent. The two conditions—sickle-cell anemia (*SS*) and sickle-cell trait (*AS*)—are often collectively referred to as *sickle-cell disease*.

The reason the *S* variant is more common in people with African ancestry has to do with recent human evolution and malaria, a devastating infectious disease caused by a microscopic parasite transmitted into the blood through mosquito bites. There are several types of malaria, each caused by a slightly different species of the parasite. The parasite species *Plasmodium falciparum* causes the most severe form of malaria, and nowhere in the world is this type of severe malaria more prevalent than in tropical Africa.

To best understand the relationship between severe malaria and sickle-cell trait, we need to go back more than two million years in human evolutionary history. There were no anatomically modern humans at that time, and our humanlike ancestors lived exclusively in Africa. They were susceptible to a milder form of malaria—not the severest form that now plagues humans, but nonetheless serious enough to cause illness and death. Chimpanzees continue to suffer from this milder form, but humans today are immune to it. A mutation in a different gene (*CMAH*) in one of our ancient ancestors produced a variant that conferred complete immunity to this milder form of malaria. Natural selection favored this variant in our ancestors until it eventually was the only variant present in ancient humans, the original ancestral variant having disappeared entirely. All humans alive today are homozygous for this protective variant in the *CMAH* gene and are immune to the form of malaria that still infects chimpanzees.

However, immunity to malaria would not last. The parasite that causes malaria is a living organism with its own DNA, so variants within its DNA that increase its ability to infect humans may be favored through natural selection. Over a period of nearly two million years, several variants accumulated in the ancient malarial parasite's DNA that allowed a new species to evolve, one that infects only humans: *Plasmodium falciparum*. This new species overcame the genetic immunity that had protected our ancestors. The final mutation in the parasite happened quite recently, between five thousand and ten thousand years ago, after humans had spread throughout the world.[2] This newly evolved parasite now causes the severest form of malaria. It is exclusive to humans, and it is the most common type of malarial infection today.

The *S* variant confers resistance to infection by this parasite, especially in young children, so people who carry one copy of the *S* variant (*AS*)—in other words, who have sickle-cell trait—are more resistant to malaria than those who carry two copies of the ancestral variant (*AA*). In temperate regions of the world where malaria is absent or rare, the *S* variant confers little or no advantage. In fact, throughout most of human history, it was *disadvantageous* in temperate regions. Without the modern medical treatments now available, children who had sickle-cell anemia (*SS*) often died before they could reproduce and pass on the *S* variant to children. In regions such as equatorial Africa, however, where malaria is endemic, people with sickle-cell trait (*AS*) had an advantage for survival and reproduction because they were resistant to malaria. And their offspring who inherited one copy of the *S* variant (*AS*) also had the same advantage.

Throughout several thousand years in Africa, and, to a lesser extent, in the Arabian Peninsula, south Asia, and the Mediterranean region, natural selection favored the *S* variant in people who had sickle-cell trait, maintaining a relatively high prevalence of the S variant. Hence, people today whose ancestry derives from regions where malaria historically was prevalent are more likely to carry the S variant and have sickle-cell trait. Also, sickle-cell anemia is more common in people whose ancestries trace to these regions, although much less common than sickle-cell trait because the probability of inheriting the S variant from *both* parents (who must both be AS for this to happen) is less than inheriting it from just one.

There is solid evidence that the S variant originated several times independently in Africa—in one case, less than 2,100 years ago.[3] And the S variant present in people whose ancestry is from a region stretching from the Arabian Peninsula to south Asia originated from yet another independent mutation.[4] Each time the S variant arose from mutation in an area where malaria was endemic, natural selection promoted an increase in its prevalence. Therefore, this variant is not purely African in origin, only more frequent in people with African ancestry. Also, importantly, not everyone with African, Arabian, or south Asian ancestry carries the S variant; in fact, just a small minority do, so having ancestry from one of these regions is no guarantee that someone carries the S variant, only a somewhat increased probability compared to people who do not have ancestry from these regions.

Sickle-cell trait is an example of how evolutionary history can become entangled in modern racial tensions. After the tragic death of the young man mentioned at the beginning of this chapter, his family discovered that he was not the first athlete to die from complications of sickle-cell trait after a strenuous practice; several African American collegiate athletes who had sickle-cell trait had previously died under similar circumstances. Quite a few people argued that had he and the others been tested and their health status made known, they probably would not have been subjected to the strenuous conditions that ended up taking their lives. His family sued Rice University and the National Collegiate Athletic Association (NCAA). Families of other athletes sued as well. Eventually, after legal settlements, the NCAA implemented rules regarding testing for sickle-cell trait. Beginning in 2010, the NCAA required all collegiate athletes, regardless of their ancestral background, to be tested for sickle-cell trait, provide evidence that they have already been tested, or sign a waiver releasing liability should they choose not to be tested.

The rule immediately sparked controversy, including stern warnings from reputable scientific organizations about its potential for racial discrimination. According to an official statement issued in 2012 by the American Society of Hematology, representing more than sixteen thousand physicians and scientists specializing in blood diseases, "the NCAA policy attributes risk imprecisely, obscures consideration of other relevant risk factors, fails to incorporate appropriate counseling, and could lead to stigmatization and racial discrimination."[5] This scientific society recommended instead "the implementation of universal interventions to reduce exertion-related injuries and deaths, since this approach can be effective for all athletes irrespective of their sickle cell status."[6] According to a statement by the Sickle Cell Disease Association of America, the current NCAA requirement "carries great risk of stigmatization and discrimination against athletes with sickle cell trait. The NCAA mandate for sickle trait screening does not provide adequate assurance of the privacy of genetic information nor protection from the discriminatory use of such information."[7]

The US Army had faced a similar situation. Research on the rare deaths of recruits during basic military training revealed a higher incidence of deaths for those who had sickle-cell trait than for those who did not.[8] Notably, most recruits with sickle-cell trait passed basic training without incident, and a few who did not have sickle-cell trait also died from overexertion. Sickle-cell trait merely increased the statistical likelihood of death due to overexertion. In response, the army altered its protocols for basic training to better protect *all* soldiers from risks associated with overexertion, heat exhaustion, and dehydration, regardless of their sickle-cell status. In fact, no branch of the US military currently tests for, or requests information about, a recruit's sickle-cell status. Large numbers of Brazilian citizens also have African ancestry, with a somewhat higher incidence of sickle-cell trait. The Brazilian Army has implemented a similar policy to protect all soldiers from overexertion and does not test them for sickle-cell trait. The American Society of Hematology has recommended that collegiate athletic programs do the same.

Sickle-cell trait is perhaps the best known of many genetic conditions that are correlated with ancestry. As another example, malaria is also common in parts of Southeast Asia, yet sickle-cell trait is less common in these regions. In this part of the world, natural selection has favored a different set of derived variants that confer resistance to malaria, variants that do not cause sickle-cell trait or sickle-cell anemia. Instead, when homozygous, several of these variants cause serious, often fatal, blood-related conditions called *thalassemias*. Symptoms include extreme weakness and a tendency to become tired very quickly, jaundice (skin yellowing), swollen abdomen, bone deformities, and delayed growth. A person who is heterozygous, having inherited just one copy of one of these variants, however, has increased resistance to malaria and does not have symptoms of thalassemia. Because of natural selection, thalassemias are some of the most common inherited conditions in people whose ancestry is from regions of Asia plagued by malaria, especially Southeast Asia. Thalassemias, however, are far from exclusive to people with Southeast Asian ancestry. They are due to a large number of different variants found in human populations throughout the world. The prevalence, however, is greatest in people whose ancestry traces to regions with malaria, due to natural selection favoring resistance to malaria in their ancestors.

An additional example of a frequent and serious genetic condition associated with geographic ancestry is cystic fibrosis. The symptoms of this condition vary among people who have it; they often include thickened mucus that causes chronic coughing and wheezing, susceptibility to lung infections, and digestive problems. Derived variants in a gene called *CFTR* cause it, and a person must inherit these derived variants from both parents (in other words, must be homozygous) to have cystic fibrosis. Those who have just one derived variant are called *heterozygous carriers*, and they may pass the variant on to their children, but they have no symptoms. The majority of cystic fibrosis cases result from homozygosity for the same derived variant, called *delta F508*, that lacks three base pairs when compared to the ancestral variant:

Ancestral variant
```
AAAATATCATCTTTGGTGTTT
TTTTATAGTAGAAACCACAAA
        deletion
AAAATATCATTGGTGTTT
TTTTATAGTAACCACAAA
```
Derived variant

The word *delta* here stands for "deletion," meaning that a mutation deleted three base pairs to create the derived variant. Evidence in the DNA surrounding this variant indicates that its originating mutation happened only once in human history, so everyone who inherits this variant, be it one copy or two, traces her or his ancestry for this part of their DNA to the same ancient person. This variant is so ancient that it is difficult to trace where this person lived. He or she probably lived somewhere between central Asia and Europe more than thirty thousand years ago, and the distant descendants of this person colonized much of Europe generations later.[9] The relatively high proportion of people with European ancestry who carry this variant is probably a consequence of natural selection. Heterozygous carriers of the variant are less likely to become severely dehydrated by diarrhea; thus, they are better able to survive cholera infections and possibly more resistant to typhoid fever. This survival advantage probably hastened the spread of this variant and contributed to its prevalence in ancient Europe. Hence, cystic fibrosis appears most frequently (albeit not exclusively) in people who have European ancestry, and, for this reason, it is the most common genetic disorder in Europe, Canada, Australia, and the United States, as well as other countries with high proportions of people who have predominantly European ancestry.

All of these examples—sickle-cell trait, sickle-cell anemia, thalassemias, and cystic fibrosis—are genetic conditions caused by identifiable derived variants. The causal relationship of each of these conditions is clear and absolute, understood in detail at the molecular, cellular, and physiological levels. And we could examine many similar examples.

More difficult to identify, however, are variants that do not necessarily *cause* a condition or disease but, rather, confer *susceptibility* to it. For example, type 2 diabetes, heart disease, obesity, rheumatoid arthritis, Alzheimer's disease, Parkinson's disease, autism, various types of cancer, and many other health-related conditions tend to run in families but do not display clear-cut patterns of inheritance like cystic fibrosis or sickle-cell anemia do. Some of these conditions, like type 2 diabetes, heart disease, obesity, and types of cancer, are not entirely genetic but also dependent on diet, physical activity, exposure to chemical substances and environmental pollutants, and other non-genetic factors—an interaction of inheritance and environment.

In science, it is often a challenge to distinguish causation from mere correlation. Although correlation is often a result of causation, a cardinal rule of science is that correlation in and of itself does not provide sufficient evidence for causation. As a simple and obvious example, Utah has one of the highest rates in the United States of malignant melanoma, the most dangerous and lethal form of skin cancer.[10] Utah also has the lowest rate of tobacco smoking.[11] Yet no legitimate scientist would recommend that people take up smoking to prevent malignant melanoma because there is no scientific reason to presume any causal relationship between the two. Indeed, if there were any causal relationship, it should be the reverse.

By contrast, the causal relationships for many genetic associations are evident based on a clear understanding of how genes function and how variants in them alter their functions. If a variant causes a change in the function of a gene related to a particular condition, and a genetic variant in that gene is correlated with the incidence of that condition, the correlation probably represents at least a partial causal relationship between the two. We can extend the example we just examined—the relatively high incidence of melanoma in Utah—as a case in point. Utah has the second-highest rate of malignant melanoma, and Vermont has the highest. These two states have relatively high proportions of people with northern European ancestry (in fact, Vermont's is the highest in the United States). Utah also has relatively high ultraviolet light radiation, due to the high elevation where most of its population lives (an environmental risk). As we've already seen, people with predominantly northern European ancestry typically carry derived variants that confer low skin pigmentation, which increases their susceptibility to malignant melanoma when exposed to sunlight. Therefore, susceptibility to malignant melanoma and variants that cause reduced skin pigmentation *are* causally correlated. This same correlation is evident on the other end of the spectrum. In the United States, malignant melanoma is lowest within the District of Columbia (Washington, DC), which has the highest proportion of people with predominantly African ancestry and high skin pigmentation, and, therefore, ancestral variants that confer protection against this type of cancer.

Correlations between diseases and the variants conferring susceptibility to them are not absolute. Malignant melanoma is an example. People who are genetically susceptible to it because of lower skin pigmentation can reduce the probability of malignant melanoma by avoiding overexposure to the sun and protecting their skin with clothing and sunscreen.

Among the most researched genetic susceptibilities associated with ancestry is predisposition to alcohol dependence, popularly known as alcoholism, which affects large numbers of people of all ancestral backgrounds in every part of the world. Humans, as well as many other species of animals, have genes whose products metabolize alcohol to protect them from its harmful effects (in excessive amounts, alcohol is toxic, even fatal, to humans). These gene products act mostly in the liver, which is why people who consume large amounts of alcohol over long periods during their lifetimes often suffer from liver disease. When someone consumes alcoholic beverages, the alcohol is quickly absorbed into the bloodstream and soon affects the brain and nervous system. It initially causes sensations of well-being, including the so-called buzz, followed by drunkenness with increased consumption. The liver metabolizes the alcohol through a two-step process, beginning with conversion of ethanol (the type of alcohol in alcoholic beverages) to an intermediate substance called acetaldehyde. The second step is converting acetaldehyde to acetate, a substance that causes the sour taste in vinegar. In fact, the process that happens in the liver is similar to the chemical process that naturally converts wine to vinegar when wine is exposed to oxygen in the air.

Variants in the genes that govern this process may alter the body's ability to metabolize alcohol. Which combination of these variants a person carries may either increase or decrease that person's susceptibility to alcohol dependence. The genetic condition is a case of susceptibility because alcohol dependence is not entirely genetic but also influenced by behavior. Obviously, strict abstinence from alcohol consumption prevents alcohol dependence regardless of which genetic variants a person carries. But for those who regularly consume alcohol, susceptibility to alcohol dependence may vary depending on genetic constitution, as well as on how often and how much alcohol they consume.

As a specific example, certain combinations of variants in genes called *ADH1B*, *ADH1C*, and *ALDH2* cause acetaldehyde, the product of the first step, to accumulate to excessive levels when a person consumes alcohol. Some variants increase the rate of the first step (conversion of ethanol to acetalde-hyde); others reduce the rate of the second step (conversion of acetaldehyde to acetate). In either case, the result is similar: acetaldehyde accumulates to high levels, causing a reaction named *facial flushing syndrome* or *alcohol flush reaction*. The word *syndrome* is appropriate because the combination and degree of symptoms often vary, depending on the individual's genetic constitution and the amount of alcohol consumed. Symptoms may include facial reddening, redness in other parts of the body, nausea, headache, confusion, dizziness, and blurred vision, often with a reduction in the pleasant effects that come with the initial stages of inebriation. People who suffer from facial flushing syndrome after alcohol consumption are often hesitant to consume large amounts of alcohol, so the syndrome decreases their susceptibility to alcohol dependence. Some medications used to treat alcohol dependence cause facial flushing syndrome, creating an unpleasant sensation when alcohol is consumed as they inhibit conversion of acetaldehyde to acetate.

Some common terms for facial flushing syndrome are "Asian glow," "Asian blush," and "Asian flush" because this syndrome is more common in people with ancestry from east Asia, especially eastern China, Japan, and the Korean Peninsula. Like many other genetic conditions, it is not exclusive to people with a certain ancestry, simply more frequent among them—a fact that led the author of a popular article in *Yale Scientific* to conclude: "Maybe it's time, then, to think of a new name for 'Asian glow.'"[12]

Several of the variants that cause the syndrome originated anciently in east Asian populations and have spread among their descendants. One of the best-studied examples is a derived variant in the *ADH1B* gene. This variant increases the conversion of ethanol to acetaldehyde by about a hundredfold compared to the ancestral variant.[13] It is unusual because most derived variants reduce or eliminate the function in the products they encode when compared to the ancestral variant, whereas this one *increases* its function. Like many of the examples we've already discussed, it originated from a single base-pair change:

Ancestral variant
GGAATCTGTC**G**CACAGATGAC
CCTTAGACAG**C**GTGTCTACTG

Derived variant
GGAATCTGTC**A**CACAGATGAC
CCTTAGACAG**T**GTGTCTACTG

The original geographic distribution of this variant is very distinct, abruptly changing along an ancient north-south linguistic and cultural divide in east Asia. The region where it is most prevalent is on the eastern side of this divide, including what is now eastern China, the Korean Peninsula, and Japan.[14] People who carry this variant are susceptible to facial flush syndrome. A second variant in this gene, which arose against the background of this first variant, is found in localized populations in extreme eastern China, the Korean Peninsula, and Japan. It further exacerbates facial flush syndrome in people who carry both variants. And there is evidence in DNA that natural selection has favored these two variants together, perhaps because they strongly inhibit alcohol dependence.[15]

Both of these variants arose in the east Asian ancestors of the linguistic and ethnic groups that occupied this region in ancient times. Geographic and cultural restrictions on mating account for the fact that these two variants are largely associated with historic ethnic populations in well-defined parts of east Asia. These restrictions inhibited, but did not entirely prevent, the spread of these variants beyond the cultural boundaries of these populations. Because of emigration during modern times, these same variants are now present in many people elsewhere in the world who have ancestry from eastern China, Japan, and the Korean Peninsula.

Associations of alcohol dependence with several different variants in the genes that govern alcohol detoxification are apparent in different populations throughout the world, detected in people with Asian, African, European, Middle Eastern, and Native American ancestry, among others.[16] Considerable attention has been directed toward Native Americans who reside on reservations in the United States, where alcohol dependence has been an extremely serious issue ever since reservations were created, afflicting the majority, and often the vast majority, of adults who live on reservations.

News stories, public protests, a documentary film, a class-action lawsuit, blogs, and books have made Whiteclay, Nebraska, infamous for alcohol abuse. As is the case on several reservations, alcohol cannot be sold on the Pine Ridge Indian Reservation, which is located immediately north of Whiteclay across the Nebraska–South Dakota border. Chris Hedges and Joe Sacco, in their book *Days of Destruction, Days of Revolt*, offer a stark description of the town:

> Whiteclay, an unincorporated village that exists for only a block and a half before vanishing into the flatlands of the surrounding prairie, has only five or six permanent residents. It exists to sell beer and malt liquor. It has no town hall, no fire department, no police department, no garbage collection, no municipal water, no town sewer system, no parks, no benches, no public restrooms, no schools, no church, no ambulance service, no civic organizations, and no library. . . . The liquor stores dispense the equivalent of 4.5 million 12-ounce cans of beer or malt liquor a year, or 13,500 cans a day. . . . Whiteclay's clients, however, are some of the poorest people in the country. They are Native Americans from the Pine Ridge reservation that is less than 200 feet away, just over the state line in South Dakota.[17]

That social issues—such as poverty, unemployment, discrimination, substandard health care, poor living conditions, and diminished educational opportunities—contribute to alcohol dependence on reservations is beyond question. However, is there evidence that genetic variants inherent to Native Americans confer increased susceptibility to alcohol dependence? Thus far, genetic analysis of several variants in alcohol-metabolizing genes among Native Americans who reside on reservations shows some statistically reliable correlations of alcohol dependence with certain variants. In some cases, the association is negative; the derived variants *protect* against alcohol dependence.[18] Other research suggests that alcohol dependence may be associated less with alcohol-metabolizing genes and more with variants in other genes that confer a generalized craving for addictive substances, including alcohol, methamphetamines, and cocaine.[19] The correlations identified thus far, however, are relatively mild. The evidence collectively indicates that, although genetic constitution may contribute some degree of either predisposition or protection, deplorable social and economic conditions on reservations are the overriding factors responsible for the rampant abuse of alcohol and other addictive substances.

Many of the variants we've examined so far arose from mutations that originated less than twenty thousand years ago in localized populations and have spread beyond their regions of origin more recently. However, even the most ancient variation that arose in Africa and is retained in people worldwide is associated with geographic differences in health.

One of the best-studied examples is the *AGT* gene, which regulates blood pressure. A derived variant is associated with lower blood pressure, which confers lower incidence of heart disease in people alive today. This variant is very old, having first appeared in Africa before the ancient out-of-Africa migrations. Both variants are now present in populations throughout the world but are distributed unevenly. The ancestral variant is more common in Africa, as well as in some regions of the world outside Africa. The derived variant bears the marks of a selective sweep in several places where it is more prevalent, outside Africa.[20] One possible explanation for the uneven distribution is natural selection for salt regulation. This gene regulates the amount of salt retained by the body, and the ancestral variant tends to cause salt retention, a trait that was advantageous in parts of the world where dietary salt was in short supply, which was the case in much of Africa. Other parts of the world have abundant salt, however, and in these regions, salt retention can be a liability because excess salt increases blood pressure. Ancient diets probably varied in the amount of salt they had, resulting in different effects of natural selection.

These examples of health issues are just a very small sampling of hundreds associated with variants in DNA dispersed among the world's people. Ancient African variants as well as more recent variants are distributed unevenly throughout the world's human population, and the incidence of diseases influenced by these variants is often correlated with geographic ancestry. A combination of factors is responsible for this nonuniform distribution of variants, including where and when mutations originated, emigration and settlement patterns of ancient humans, historic mating and cultural practices, geographic barriers and topography, climate, random fluctuations in variant frequencies from one generation to the next, and the influence of natural selection.

Throughout much of human history—including recent history, right up to the present—false assumptions about the relationship of race, ancestry, health, and inheritance have abounded. In some cases, racist misconceptions reinforced these false assumptions. Perhaps nowhere is this more evident than with the history of sickle-cell disease in the United States. Sickle-cell anemia has been known as a specific disease for little more than a century, first identified in 1910 in an African Caribbean young man who was a student in the United States.[21] In the ensuing years, additional cases of sickled cells in blood were identified, some associated with outward symptoms of sickle-cell anemia but most without symptoms (the genetic distinction between sickle-cell trait and anemia remained unknown until 1949), which contributed to a plethora of false assumptions prior to that time.[22] According to two physicians writing in 1930 about what later was identified as sickle-cell trait, "the sickler who presents even mild anemia is a subnormal individual and even though he may not be regarded as an active case of sickle cell anemia, he is still ill equipped to withstand the vicissitudes of life."[23]

At first, all cases of sickled cells were identified exclusively among African Americans, and the disease became widely characterized as a "Negro disease." Although a few cases were identified in people of European, Middle Eastern, and south Asian ancestry, they were often attributed to supposed undocumented African ancestry under the flawed assumption that the genetic factor causing sickled cells could only be African in origin. By the 1940s, scientists were in the midst of a concerted effort to determine the extent of cell sickling in Africa. They found high prevalence of people with sickled cells but apparently few cases of sickle-cell anemia. Most physicians working in Africa ascribed this observation to three factors: 1) inadequate diagnosis, 2) high mortality rates in Africa for children with sickle-cell anemia, and 3) difficulties distinguishing the symptoms of sickle-cell anemia from those of malaria. However, others attributed the supposed higher prevalence of sickle-cell anemia in African Americans to so-called racial mixing in their ancestry.

It was well known at the time that most African Americans had some European ancestry, as is now well documented by modern genetic evidence.[24] This was, in large part, a result of sexual abuse by slave owners and masters, abundantly recorded in the dictated recollections of former slaves.[25] African Americans, therefore, were viewed during the middle of the twentieth century as a mixed or hybrid race, as opposed to native people who resided in Africa. In 1950, A. B. Raper published an extensive review of the scientific literature available at the time. Referring to this supposed higher incidence of sickle-cell anemia in African Americans when compared to native Africans, he wrote that "some factor imported by marriage with white persons, is especially liable to bring out the haemolytic [anemic] aspect of the disease, while the anomaly remains a harmless one in the communities in which it originated."[26] This view, widely held at the time, supposedly justified the assertion that African Americans were members of a genetically inferior mixed race, inferior both to "pure" Europeans and "pure" Africans. It further lent erroneous support to the notion of white supremacy, and laws mandating antimiscegenation and racial segregation.

Scientific research later dispelled these fallacies. Sickle-cell anemia was indeed present in Africa, with symptoms as severe as elsewhere. It also appeared in people with no African ancestry, particularly in the Arabian Peninsula, south Asia, and the Mediterranean region, areas where malaria was prevalent. And the notion of so-called pure races had no support in genetic data. Nonetheless, dogmatically held opinions regarding white supremacy, the supposed inherent inferiority of African Americans, and the presumption that racial purity was essential remained strong well into the 1960s and '70s, and even to the present, often fallaciously supported through inaccurate suppositions about sickle-cell disease.

As the civil rights movement gained momentum in the late 1960s, political agendas to improve racial equality in the United States emphasized sickle-cell anemia as a priority for research and treatment. Prior to that time, genetic conditions that were more common in European Americans had received the lion's share of governmental and philanthropic research funding. Robert B. Scott, a physician and professor at the Howard University College of Medicine in Washington, DC, was one of the strongest and most vocal advocates for increasing research funding on the biological basis of and treatment for sickle-cell anemia. In an influential 1970 article, he lamented the broad neglect of sickle-cell anemia and the lack of funding to support research and treatment:

In 1967 there were an estimated 1,155 new cases of SCA [sickle-cell anemia], 1,206 of cystic fibrosis, 813 of muscular dystrophy, and 350 of phenylketon- uria. Yet volunteer organizations raised $1.9 million for cystic fibrosis, $7.9 million for muscular dystrophy, but less than $100,000 for SCA. National Institutes of Health grants for many less common hereditary illnesses exceed those for SCA.[27]

Underfunding for sickle-cell anemia was more a political issue than a scientific or medical one. Sickle-cell anemia was medically more serious than many less prevalent diseases yet was not a high priority for government or philanthropic support. According to Melbourne Tapper, writing in retrospect in 1999, "African Americans sought to increase funding for sickling research by turning to telethons, modeled on those for cystic fibrosis, muscular dys- trophy, and cerebral palsy. These telethons were unsuccessful not because of the clinical nature of sickling, but because they were unable to neutralize the historical difference of the population in which sickling was primarily found—African Americans."[28]

A biochemical test that could accurately diagnose both sickle-cell anemia and sickle-cell trait had been available since 1949.[29] Several states imple- mented testing programs—voluntary in some cases, mandatory in others— targeting the African American population, often with well-documented instances of racism in the administration of these tests. Against this back- drop, President Richard M. Nixon proposed increased funding for sickle-cell research, and Congress responded by passing the National Sickle Cell Anemia Control Act (the word *Control* was later changed to *Prevention*), which Nixon signed into law in 1972. Although testing was encouraged, it was voluntary, partially overcoming some of the earlier claims of racist coercion associated with mandatory testing.

The purpose of testing was to inform potential parents who both were heterozygous carriers (in other words, who both had sickle-cell trait) of the possibility of having a child with sickle-cell anemia. As envisioned by Scott, "Whether a young couple will decide to have no children, or plan a limited family size, or disregard the risk would be entirely their own decision."[30]

Despite its lofty goals, the impact of this act quickly faded. Congress failed to appropriate sufficient funding for it, and the act expired three years after it was signed into law. Several clinics established under it had to be closed for lack of funds. Federal funding for sickle-cell anemia was later incorporated into funding for genetic diseases in general, so it once again had to compete with diseases that were most prevalent among European Americans, such as cystic fibrosis and muscular dystrophy.[31]

Research in 1986 showed that early intervention with treatments for infants with sickle-cell anemia could significantly improve their lifelong outlook for health. This finding offered an impetus for mandatory newborn screening. Newborns identified with sickle-cell anemia could be immediately identified and provided treatment, thereby increasing their lifetime outlook for health. Over the next twenty years, states began implementing sickle-cell testing for newborns, with universal testing in all fifty states and the District of Columbia by 2006. With the support of sickle-cell organizations, healthcare professionals, and the National Association for the Advancement of Colored People (NAACP), Congress passed the Sickle Cell Treatment Act of 2003, which President George W. Bush signed into law. The act provided federal funding for research, counseling, education, matching funds for Medicaid to assist with treatment, and establishment of sickle-cell centers throughout the United States.

Sickle-cell disease is, without doubt, the most prominent example of how health, inheritance, and ancestry have become entangled with racial tensions. Those tensions have persisted for more than a century and are still with us, as the latest controversy regarding testing athletes for sickle-cell trait attests. There are other examples as well. Although perhaps not as well known, they, too, illustrate how ignorance of scientific information can result in discrimination, whether intended or not.

An example is lactose intolerance—the inability to fully digest dairy products, especially fresh milk—which is common throughout the world, affecting more than 65 percent of the world's population. In fact, it is the original ancestral state of humanity. Mammals, including humans, consume milk during infancy, then are weaned from milk as they begin consuming other foods. The principal sugar in milk is lactose, and the body must break it down into other sugars to digest it. A single gene in our DNA, called *LCT*, encodes a protein called lactase, which carries out the first step of lactose metabolism. This gene is active during infancy but, in many people, is genetically programmed to shut down after weaning because anciently, before humans domesticated milk-producing animals, the gene was no longer needed in children who were weaned. Some people carry a derived variant that disrupts this shutdown, retaining *LCT* gene activity into adulthood and allowing them to continue consuming milk, a condition known as *lactase persistence*. Several derived variants that confer lactase persistence have arisen independently in humans, and they are mostly found in people whose ancestry traces to populations that relied on domesticated animals for milk, such as cattle, sheep, and goats.

There is good evidence that these variants were strongly favored through natural selection in people from regions where domesticated milk-producing animals were raised. Milk and other dairy products are highly nutritious, an excellent source of calories, vitamins, and minerals, especially when food is in short supply, as it often was during ancient times. In regions where humans used domestic animals for milk, people who could consume milk and milk products as a source of food had an advantage for survival and reproduction over those who could not. Natural selection favored these variants, rapidly increasing their prevalence in milk-consuming societies, and this evolutionary pattern has repeated itself independently in several parts of the world.

For instance, in East Africa, in what is now Kenya and Tanzania, nomadic herders began using domestic animals for milk more than seven thousand years ago. A large proportion of their modern descendants, most of them still in Africa, carry a specific variant that allows the *LCT* gene to remain active into adulthood.[32] In the Arabian Peninsula, where milk use was and is common, a different variant conferred lactose persistence to people who lived there anciently and their modern descendants. Yet another variant that confers lactose persistence is common in people whose ancestry is northern European, where milk from cows and goats has long been used as a source of food. This variant is very common in North America, present in approximately 77 percent of North Americans whose ancestry is predominantly European, and accounts for the high consumption of dairy products in Europe, the United States, Canada, Australia, and other parts of the world where large proportions of people have European ancestry.

Ancient Native Americans, however, never domesticated animals for milk production. Not surprisingly, the variants that confer lactose persistence in other parts of the world are rare in Native Americans, who typically begin losing the ability to digest milk by age three. Ignorant of the high proportion of lactose-intolerant people among Native Americans, and the underlying science, officials promoting US government food assistance programs distributed surplus milk products to people living on Indian reservations, where the products made most people sick. Shirley Hill Witt, a Native American anthropologist and administrator for the US Commission on Civil Rights, described the situation on a Navajo reservation this way:

> What is good for the Anglo body may not in fact be good for everyone else. This may be another mindless prejudice yet to be purged: *nutritional ethnocentrism.* To put it another way, the consequences of ethnocentrism may be more tenacious and deep-seated than we have thought. In the animal pens near Navajo hogans you can usually find the remains of milk products from the commodities program: butter, cheese, dried milk.

> But as more and more investigations are reported, the fact is becoming incontrovertible that for many or most of the world's people, milk is not our most valuable food, or "nature's way," or so say the slogans of the milk industry. These studies indicate that most of us cannot drink milk after early childhood without suffering gastric upset, cramps, bloating, diarrhea and nausea.[33]

Promotion of milk and other dairy products for consumption by children in public school cafeterias likewise ignores the pervasive nature of lactose intolerance among many children who do not descend from ancient milk-reliant cultures. This is especially relevant in schools located on or near reservations. According to Witt,

> In schools across the nation, children are browbeaten into ingesting vast quantities of milk whether or not they have the genetic equipment to do so. In 1972, a study I conducted in one of the New Mexican pueblos showed that only one person out of a hundred over the age of six was able to tolerate lactose without strong digestive reactions.[34]

Reliable laboratory tests for lactose intolerance are available for administration under the supervision of a physician. Most are not genetic tests but, rather, tests that directly measure a person's ability to digest lactose. For those who have lactose intolerance, lactose-reduced and lactose-free dairy products are available, as are supplements that assist the body with lactose digestion, allowing a larger number of people with lactose intolerance who desire to consume dairy products to do so.

Because ancestry is closely tied to a wide variety of health issues, physicians have historically used racial categories to recommend different tests and treatments. The intent is not racial bias but, rather, a means to more efficiently direct medical interventions according to information published in the medical literature about health issues and ancestry. For instance, targeting diagnosis and genetic testing of cystic fibrosis in people with predominantly European ancestry, or sickle-cell disease in people with significant African ancestry, made economic sense to many healthcare organizations.

Simple blood tests that are inexpensive, rapid, and easy to administer can readily detect as many as twenty-nine genetic conditions in infants. Although some of the conditions they detect are more common in people whose ancestries trace to particular parts of the world, the American College of Medical Genetics has recommended *universal* screening of infants for all twenty-nine of these conditions, rather than targeted screening by ethnic group. The reasons are to avoid missed diagnoses and to treat these conditions in time to avert the most serious consequences associated with them.[35] Targeting by ethnic group inevitably misses cases because the ethnic classifications of infants are not accurate assessments of ancestry. Moreover, the history of targeted screening has shown that discrimination and stigmatization are unavoidable consequences, whether intended or not. There is no valid medical reason to consider racial or ethnic classification for such testing.

Until 2005, no medication had been approved by the US Food and Drug Administration (FDA) for treatment of any particular ethnic group. That year, a drug known as BiDil was approved in the United States for treatment of congestive heart failure specifically in African Americans, based on the results of clinical trials. An initial clinical trial for BiDil had included people of different ancestral backgrounds. The results initially showed little advantage for the drug until the researchers revisited the data, parsing the analysis according to self-identified racial classification. They then discovered a possible benefit for subjects who self-identified as African American. This prompted another clinical trial with only African American participants. All participants were already suffering from congestive heart failure at the beginning of the study and were on other medications to treat their condition. The researchers randomly assigned each of them to receive BiDil or a placebo in addition to the medications each was currently taking. The trial was to continue for eighteen months but was terminated early because those taking BiDil had a lower rate of death. The FDA approved BiDil for use in African American patients in 2005 on the basis of this trial.

Some groups, such as the NAACP, praised this action for focusing medical research on an ethnic group that had long suffered medical discrimination. In fact, medical research targeting African Americans has an appalling historical record, exemplified by the unsubstantiated claims and outright errors made throughout the twentieth century with sickle-cell disease. Perhaps the most infamous case, however, was research conducted from 1932 through 1972 in Tuskegee, Alabama, in which African American men were misled into enrolling as participants in a study purportedly about blood disorders. The real purpose of the study, which was kept secret from the participants, was to research syphilis. The majority of the men enrolled in the study already had the disease when the study began, and others were intentionally infected without their knowledge. None were told that they had syphilis, and none were treated for it, even though penicillin was found to be an effective treatment during the early years of the study. Not only was syphilis allowed to progress unabated in these men, many of their spouses became infected, as did infants born to these women. The study was finally terminated after forty years when a whistleblower took the story to the press after being rebuffed when he reported its abuses to the responsible government agencies. In the aftermath, Congress mandated substantial changes in legal and ethical requirements for government-sponsored research.

Clinical trials for BiDil, with their exclusive focus on African Americans, seemed to offer a hint of reversal after decades of past injustices. According to company officials, "BiDil was 'the antithesis of Tuskegee'" and "the approval of BiDil was about putting Tuskegee to rest."[36] However, geneticists, medical researchers, ethicists, legal experts, and a considerable number of physicians criticized the testing and release of BiDil as a marketing strategy carefully crafted to generate corporate profits. The drug was a combination of two drugs already available in generic forms, but the dose used in the study could not be easily formulated with available doses of generic alternatives. Had physicians been able to readily prescribe a generic alternative, it could have been equally effective and much less expensive. There was no research to indicate whether other doses, including those generically available, were less, more, or equally effective.

As it turned out, BiDil failed to realize market expectations. Projected prescriptions and sales to African American patients did not materialize, and the marketing campaign was disbanded in 2008. The company that marketed the drug was forced to downsize and, in 2009, was sold to another company.[37]

The drug's focus on African Americans implied a genetic basis for its effectiveness divided along supposed racial boundaries. Prominent geneticists blasted this implication, arguing that if variation in the efficacy of BiDil had a genetic basis, research identifying associations between the drug's efficacy and specific variants in DNA would be a much more reliable way to target those who would most benefit, rather than using self-identified racial classification as the criterion. According to J. Craig Venter, one of the foremost scientists and business leaders in genome-based medicine, and his colleagues, "to attain truly personalized medicine, the scientific community must aim to elucidate the genetic and environmental factors that contribute to drug reactions and not be satisfied with a race-based approach."[38] As Howard Brody and Linda M. Hunt of Michigan State University point out, there were financial disincentives that discouraged a for-profit enterprise from doing so:

> To what extent would the identification of a specific genetic trait, correlated with positive therapeutic response, be likely to expand that market? As long as there is some probability that the results of that further research could cause the market to shrink, even if by a small amount, there is every incentive for the company to decline to undertake that research.[39]

Brody and Hunt further argue that self-identified racial categories can be considered valid medical criteria not on the basis of genetics but, rather, for social and cultural reasons:

> Family physicians, well schooled in the biopsychosocial model of health, ought especially to be concerned about an approach to research that de-emphasizes the search for social and cultural factors in disease. . . .
>
> For example, hypertension, one of the major risk factors for congestive heart failure, is more common within the African American community; and chronic social stress has been implicated as a possible contributor to the development of hypertension. Diet, exercise, and other environmental variables are also possible mediators.[40]

In the end, BiDil turned out to be another instance of controversial attempts at racially based medicine. Although touted as the first case of *personalized medicine*—meaning genetically targeted treatments and interventions—in reality, it was not. Truly personalized medicine associates treatments with specific variants that can be directly identified in a person's DNA irrespective of racial classification.

As a more recent example, a June 2014 article with the title "Differing Effects of Metformin on Glycemic Control by Race-Ethnicity" received considerable attention in the popular press. It shows a more pronounced benefit of the diabetes medication metformin for African Americans than for European Americans based on self-identification. Although readers might assume the cause for this difference is genetic, the authors of the article point out in its concluding sentence that any such assumption is thus far inconclusive: "studies assessing the effect of genetic ancestry, rather than self-reported race-ethnicity, may help clarify whether there is a heritable component to population group differences in metformin response."[41]

For years, the majority of DNA tests available for use in medicine targeted single variants, and the tests were expensive, often not covered by insurance. But this scenario has dramatically changed. The cost of testing for DNA variants has fallen so precipitously that it is now readily affordable to large numbers of people. For example, tests that simultaneously detect thousands of variants currently cost less than $100 and are available without a physician's prescription. Several companies offer such tests via the Internet. A person simply makes the order, paying by credit card, and a kit arrives in the mail. DNA is provided simply by swabbing the inside of the cheek or spitting into a tube, and the sample is returned by mail to the company.

After the analysis is complete, the test results are made privately available to that person online. Ancestry informative markers reveal the most probable composition of the individual's ancestry, which often is not exclusive to a particular region but is a mix of DNA segments inherited from different parts of the world. Some tests also reveal variants associated with health, including those strictly associated with genetic conditions such as cystic fibrosis, sickle-cell disease, lactose intolerance, and numerous others. The tests may also reveal statistical associations with genetic variants that confer greater-than-average susceptibility to, or protection against, other conditions, such as type 2 diabetes, alcohol dependence, coronary heart disease, Alzheimer's disease, Parkinson's disease, susceptibility to various types of cancer, and many others. The FDA, however, issued strict rules that substantially limit what health information a company can directly provide to consumers based on DNA variants, and the rulings may end up being challenged in court.[42] The raw data, including all tested variants, remain available so that health professionals who are knowledgeable regarding associations of health with DNA variants can interpret the results. Unfortunately, many physicians are not sufficiently versed in the details of how specific DNA variants are associated with susceptibilities to provide accurate information. Instead, to obtain accurate and current information, patients must consult with clinical geneticists—physicians who are specialized in medical genetics.

Such testing is rapidly opening the door to true personalized medicine: treatments and interventions fine-tuned to one's genetic constitution. This approach can, at least in theory, make medicine more successful and cost-effective. Currently, the focus of genetic testing is on early diagnosis and intervention rather than specific fine-tuning of medications. By identifying predispositions to diseases before they become serious, and implementing enhanced screening for early diagnosis and preventative measures, such information can help people and their physicians to effectively manage healthcare. For example, a physician may determine that someone who carries a variant conferring increased susceptibility to colon cancer should have colonoscopies earlier and more frequently than those who have average or lower-than-average genetic predispositions to colon cancer. Famously, a number of women who discovered through DNA tests that they carry variants conferring increased susceptibility to breast cancer have chosen to undergo preventative mastectomies.[43]

People often fear genetic testing because of its potential for abuse, and history is replete with instances when genetic information was used to deny or restrict employment or health insurance.[44] When the Human Genome Project officially began in 1990, its leaders projected that one of many offshoots of the project would be a dramatic increase in genetic testing. Recognizing how serious discrimination on the basis of genetic tests had been in the past, they formed a committee to study the ethical, legal, and social issues and devoted 5 percent of the project's funding to this effort. Among the several recommendations by this committee was legal protection against genetic discrimination. The committee crafted proposals that eventually became the Genetic Information Nondiscrimination Act (GINA). In spite of broad support in both houses of Congress, the legislation languished unimplemented for thirteen years, from 1995 through 2008, in large part because of opposition from corporations and insurance companies that might experience financial losses through restrictions on their ability to manage risk. Finally, the legislation made its way through the House of Representatives, passing overwhelmingly in April 2007 by a vote of 414 to 1. After delays postponed its consideration by the Senate, it finally passed almost a year later by a vote of 95 to 0. President George W. Bush signed it into law on May 21, 2008.[45]

GINA specifically prohibits discrimination for employment or health insurance on the basis of genetic information. However, it does not exclude discrimination for other forms of insurance, such as life or disability insurance. Large numbers of people are still reluctant to undergo genetic testing out of fear that the information may be used for discrimination in spite of current legal protections.

A long-standing concern with genetic tests is the past history of and current potential for racial discrimination. Any discrimination that stigmatizes or denies opportunity to people who carry a genetic condition that is more prevalent in people with a particular ancestry inevitably constitutes a form of racial discrimination, even if indirect, because incidence of the condition and geographic ancestry may be correlated. And, as we are about to see, health is not the only issue with the potential for indirect racial discrimination. In the following chapter, we tackle the most controversial subject that has confronted the intersection of science and race: claims that intelligence differs genetically between racially defined groups.

CHAPTER 6

HUMAN DIVERSITY
AND INTELLIGENCE

In 1981, Stephen J. Gould published one of the most significant works of his illustrious career: a book titled *The Mismeasure of Man*.[1] It reviews how scholars from the eighteenth through the twentieth centuries attempted to quantify intelligence through verbal and written tests, as well as measurements of physical characteristics such as facial features or brain size. The book reproduces caricature-like drawings from the eighteenth century that depict humans from Africa as resembling gorillas and chimpanzees: oddly exaggerated heads highlight the physical characteristics that purportedly predispose people to criminality, hand-altered photographs distort the facial features of people labeled as feebleminded, and rich convolutions in the brain of a famous mathematician imply that the brain of a genius is physically distinguishable from the less convoluted brain of a person from Papua (indigenous Papuans, at the time, were considered to be savages). Gould's book makes it obvious that these historical drawings were serious attempts to portray physically measurable features as reliable indicators of intellectual superiority or inferiority. Although atrociously humorous to us now, they serve Gould's thesis as examples of how "man" has historically been "mismeasured." Gould even addresses his choice of the seemingly gender-biased term *man* for the title by reminding us that nearly all such studies done prior to the mid-twentieth century were conducted by men, and that most of those who conducted them considered the intelligence of women to be inherently inferior to that of men.

By the time Gould published this book, he was already well known as a popular Harvard professor (shortly thereafter named the Alexander Agassiz Professor of Zoology), his specialty evolutionary biology. He was famously outspoken and controversial among scientists for novel interpretations of the fossil record. For most people, however, he will long be remembered as one of the most eloquent science writers and speakers of his day. His popular books on evolution were best sellers, often with clever titles like *The Panda's Thumb*, *Hen's Teeth and Horse's Toes*, *The Flamingo's Smile*, and *Bully for Brontosaurus*. I own two copies of *The Mismeasure of Man*. One is the original 1981 hardbound version, and the other is the 1996 paperback revised and expanded edition.[2] On the cover of the latter is a prominent statement: "The definitive refutation to the argument of *The Bell Curve*."

This statement refers to the 1994 bestselling book *The Bell Curve: Intelligence and Class Structure* by Richard Herrnstein and Charles Murray. It is a thick book, more than eight hundred pages, brimming with seemingly abundant statistical detail regarding research on intelligence and its relationship to a wide range of socioeconomic, political, educational, and biological factors— among them, race.[3] As the Edgar Pierce Professor of Psychology at Harvard, Herrnstein worked not far from Gould. In spite of their physical proximity, the two could hardly have been further apart on the subject of race and human intelligence. Herrnstein had written a 1971 article in *Atlantic Monthly* titled "IQ," which addressed the issue of race and intelligence from the point of view that genetic differences between races partially determine between-race differences in average IQ scores. Gould was outspokenly opposed to this view.

This proposition—that differences in intelligence have a predominantly genetic basis—became known as *biological determinism* or *hereditarianism*. Following from it is the notion that genetic constitution largely determines social and economic status. This latter proposition is known as *social Darwinism* (a misnomer, since Darwin neither invented nor promoted it). Gould defines it as "a specific theory of class stratification within industrial societies, particularly to the idea that a permanently poor underclass consisting of genetically inferior people had precipitated down into their inevitable fate."[4] Some proponents of social Darwinism presume that white supremacy and racial segregation are natural and inevitable, a consequence of genetic predisposition for intelligence.

The response to hereditarian ideas regarding race in Herrnstein's article was vehement. According to neuropsychologist Christopher Chabris, Herrnstein's "lectures were filled with protesters, and his speeches at other universities were canceled, held under police guard, or aborted with last-second, back-door escapes into unmarked vehicles. Death threats were made."[5]

Herrnstein and Murray were not lone voices in support of hereditarianism and its social and political implications, nor were they plowing new ground. A significant proportion of material in *The Bell Curve* is based on the work of Arthur Jensen, a professor of educational psychology at the University of California, Berkeley; a prolific researcher; and an author of numerous articles and books. Although Jensen had published much on the psychology of intelligence during the 1950s and '60s, a 1969 article titled "How Much Can We Boost IQ and Scholastic Achievement?" in the *Harvard Educational Review* established much of his fame and notoriety.[6] The article's most controversial conclusion is the hereditarian claim that genetic predisposition is responsible for a major proportion of differences in intelligence—including those between racial groups—and that because of strong genetic predisposition, social programs designed to overcome such differences are destined to fail. A 2005 article that Jensen coauthored with psychology professor J. Phillipe Rushton of the University of Western Ontario (also a strong proponent of hereditarianism) aptly summarizes the main points of Jensen's 1969 article:

(a) IQ tests measure socially relevant general ability; (b) individual differences in IQ have a high heritability, at least for the White populations of the United States and Europe; (c) compensatory educational programs have proved generally ineffective in raising the IQs or school achievement of individuals or groups; (d) because social mobility is linked to ability, social class differences in IQ probably have an appreciable genetic component; and tentatively, but most controversially, (e) the mean Black–White group difference in IQ probably has some genetic component.[7]

Gould, an outspoken critic of the rising tide of hereditarianism, denounced *The Bell Curve*'s "claim for inherited racial differences in IQ—small for Asian superiority over Caucasian, but large for Caucasians over people of African descent,"[8] as well as Jensen's ideas:

This argument [in *The Bell Curve*] is as old as the study of race. The last generation's discussion centered upon the sophisticated work of Arthur

Jensen (far more elaborate and varied than anything presented in *The Bell Curve*, and therefore still a better source for grasping the argument and its fallacies). . . .⁹

From the day *The Bell Curve* hit store shelves, it fanned the flames of what already was one of the most incendiary and polemical controversies of the late twentieth century: Were racial differences for average IQ scores due largely to genetics rather than environment? Or, in other words, are some races intellectually superior to others by virtue of their genetic constitution? In line with Jensen's earlier research, *The Bell Curve* seemed to say yes, albeit a highly qualified yes.

Its publication prompted an avalanche of responses—supportive, critical, and mixed—from natural and social scientists, philosophers, reviewers, and journalists. For example, in support of Herrnstein and Murray, Christopher Caldwell of *American Spectator* wrote, "*The Bell Curve* is a comprehensive treatment of its subject, never mean-spirited or gloating. . . . Among the dozens of hostile articles that have thus far appeared, none has successfully refuted its science."[10] By contrast, Lucy Hodges, writing for *Times Higher Education*, highlighted what many critics viewed as *The Bell Curve's* promotion of social Darwinism:

> The authors [Herrnstein and Murray] recommend doing away with affirmative action on the grounds that it poisons race relations by promoting unqualified blacks. They want to drop remedial education which they say does not work and spend the money educating talented students that the economy needs. They want to change immigration policy to prevent the influx of less intelligent people, and end welfare and other government benefits which they think encourage women with low IQs to have babies."[11]

Some of the most stinging criticism was from Gould:

> The book is a manifesto of conservative ideology, and its sorry and biased treatment of data records the primary purpose—advocacy above all. The text evokes the dreary and scary drumbeat of claims associated with the conservative think tanks—reduction or elimination of welfare, ending of affirmative

action in schools and workplaces, cessation of Head Start and other forms of preschool education, cutting of programs for slowest learners and application of funds to the gifted (Lord knows I would love to see more attention paid to talented students but not at this cruel price)."[12]

Dozens of articles appeared within months of its release, and two books, *The Bell Curve Debate* and *The Bell Curve Wars*, containing collections of essays and historical context, were quickly assembled and published the following year. Other books soon followed. Polemics and misinformation abounded, prompting educational psychologist Linda Gottfredson of the University of Delaware to draft a full-page editorial published in the *Wall Street Journal* titled "Mainstream Science on Intelligence," cosigned by fifty-two of her professional colleagues (including Jensen and Rushton), although others chose not to sign when invited. It was later reprinted with a contextual history and bibliography in the journal *Intelligence*.[13]

Leaders of the American Psychological Association (APA), the foremost scientific society on psychology in the United States, determined that the debate was so pervasive and misinformed that "there was an urgent need for an authoritative report on these issues—one that all sides could use as a basis for discussion."[14] The APA Board of Scientific Affairs commissioned a task force chaired by Ulric Neisser of Emory University to prepare this report, which was published in 2006 with the title "Intelligence: Knowns and Unknowns" in the society's journal, *American Psychologist*.[15] It was widely recognized among psychologists as an authoritative review of the state of scientific understanding regarding human intelligence at the time. A substantially updated review titled "Intelligence: New Findings and Theoretical Developments" was published in 2012, also in *American Psychologist*, by a group of seven experts in the field, headed by Richard Nisbett of the University of Michigan.[16] These two reviews are among the most comprehensive summaries of the topic available.

Of the subjects treated in *The Bell Curve*, the most contentious was Herrnstein and Murray's conclusion that "the major ethnic groups in America differ, on the average, in cognitive ability," and that a portion of these differences must be attributed to genetic differences between these groups.[17] The argument in support of a hereditarian explanation for differences for intelligence among "major ethnic groups" can be distilled into a few principal points: First, most proponents of hereditarianism emphasize that the difference in average IQ scores between African Americans and European Americans is substantial. For example, a subheading in *The Bell Curve* is worded as a question: "How Large is the Black-White Difference?" The authors then respond:

> The usual answer to this question is one standard deviation. In discussing IQ tests, for example, the black mean is commonly given as 85, the white mean as 100, and the standard deviation as 15."[18]

To put this in perspective, IQ test results are normalized to an average of one hundred and a standard deviation of fifteen. The so-called "black mean" of eighty-five represents serious cognitive deficiencies in large numbers of people. Herrnstein and Murray were careful to point out that there is considerable variation in the different studies they compiled to derive this estimate. Even so, it is exactly the same estimate Jensen had arrived at twenty-five years earlier in his 1969 article. And, in recent years, several psychologists have pointed to data indicating that the gap has been narrowing, although others have marshaled different data to claim that it has not changed.[19]

Regarding Asian Americans and European Americans, Herrnstein and Murray note that the gap is much smaller: "In our judgment, the balance of the evidence supports the proposition that the overall east Asian mean is higher than the white mean. If we had to put a number on it, three IQ points currently most resembles a consensus, tentative though it still is."[20]

The crux of their argument that these differences cannot be fully explained by environment, and therefore must be partly attributable to genetic differences between ethnic groups, is best stated in their own words:

> Suppose that all the observed ethnic difference in tested intelligence originate in some mysterious environmental differences—mysterious, because we know from material already presented that socioeconomic factors cannot be much of the explanation. We further stipulate that one standard deviation (fifteen IQ points) separates American blacks and whites and that a fifth of a standard deviation (three IQ points) separates east Asians and whites. Finally, we assume that IQ is 60 percent heritable (a middle-ground esti-

mate). Given those parameters, how different would the environments for the three groups have to be in order to explain the observed difference in these scores?

... The *average* environment of blacks would have to be at the 6th percentile of the distribution of environments among whites, and the average environment of East Asians would have to be at the 63rd percentile of environments among whites, for the racial differences to be entirely environmental.

Environmental differences of this magnitude and pattern are implausible. ... An appeal to the effects of racism to explain ethnic differences also requires explaining why environments poisoned by discrimination and racism for some other groups—against the Chinese and the Jews in some regions of America, for example—have left them with higher scores than the national average.

... The heritability of individual differences in IQ does not necessarily mean that ethnic differences are also heritable. But those who think that ethnic differences are readily explained by environmental differences haven't been tough-minded enough about their own argument.[21]

After comparing arguments favoring a predominantly genetic explanation and a predominantly environmental explanation of the differences, they draw their final conclusion:

It seems highly likely to us that both genes and the environment have something to do with racial differences. What might the mix be? We are resolutely agnostic on that issue; as far as we can determine, the evidence does not yet justify an estimate.[22]

Though Herrnstein and Murray's book does not venture an estimate of how much genetic variation contributed to this difference, Rushton and Jensen's 2005 article does. They set up the controversy as a dichotomy of two contrasting models to explain average racial differences in IQ: a culture-only model, which they propose as 0 percent genetic and 100 percent environmental, and a hereditarian model, proposed as 50 percent genetic and

50 percent environmental. By the end of their article, they fully reject the culture-only model and recommended revision of their hereditarian model to 80 percent genetic and 20 percent environmental.[23] They also speculate that a Darwinian approach explains their claim of a large hereditary difference between European, Asian, and African races for intelligence and other behavioral traits:

> Evolutionary selection pressures were different in the hot savanna where Africans lived than in the cold northern regions Europeans experienced, or the even colder Arctic regions of East Asians. These ecological differences affected not only morphology but also behavior. It has been proposed that the farther north the populations migrated out of Africa, the more they encountered the cognitively demanding problems of gathering and storing food, gaining shelter, making clothes, and raising children successfully during prolonged winters. As these populations evolved into present-day Europeans and East Asians, the ecological pressures selected for larger brains, slower rates of maturation, and lower levels of testosterone—with concomitant reductions in sexual potency, aggressiveness, and impulsivity; increases in family stability, advanced planning, self-control, rule following, and longevity.[24]

These sorts of speculations—considered by many to be overtly racist and biased against African Americans—prompted a firestorm of responses. The scientific journal *American Psychologist* devoted an entire issue to the topic in 2005, titled *Genes, Race, and Psychology in the Genome Era*. The journal *Psychology, Public Policy, and Law* invited responses to Rushton and Jensen's article from prominent psychologists, responses that could hardly have been more polarized. For instance, educational psychologist Linda Gottfredson of the University of Delaware supported Rushton and Jensen's hereditarian view:

> In summary, Rushton and Jensen (2005) have presented a compelling case that their 50%–50% hereditarian hypothesis is more plausible than the culture-only hypothesis. In fact, the evidence is so consistent and so quantitatively uniform that the truth may lie closer to 70%–80% genetic, which is the within-race heritability for adults in the West. The case for culture-only

theory is so weak by comparison—so degenerated—that the burden of proof now shifts to its proponents to identify and replicate even *one* substantial, demonstrably *non*genetic influence on the Black–White mean difference in *g* [general intelligence].[25]

Richard Nisbett of the University of Michigan drew a completely opposite conclusion in his review:

> J. P. Rushton and A. R. Jensen (2005) ignore or misinterpret most of the evidence of greatest relevance to the question of heritability of the Black–White IQ gap. A dispassionate reading of the evidence on the association of IQ with degree of European ancestry for members of Black populations, convergence of Black and White IQ in recent years, alterability of Black IQ by intervention programs, and adoption studies lend no support to a hereditarian interpretation of the Black–White IQ gap. On the contrary, the evidence most relevant to the question indicates that the genetic contribution to the Black–White IQ gap is nil.[26]

Yale University professor of psychology Robert J. Sternberg (currently at Cornell University and former president of the American Psychological Association) condemned not only the scientific but also the public-policy views expressed in the article:

> J. P. Rushton and A. R. Jensen (2005) purport to show public-policy implications arising from their analysis of alleged genetic bases for group mean difference in IQ. . . . None of these implications in fact follow from any of the data they present. The risk in work such as this is that public-policy implications may come to be ideologically driven rather than data driven, and to drive the research rather than be driven by the data.[27]

The core of the hereditarian controversy as it relates to differences between racial groups centers on three major issues: 1) how races are defined and categorized as supposedly distinct genetic entities, 2) how human intelligence is measured, and 3) how those measurements are interpreted in terms of genetic and environmental causes.

The first issue—the biological and social bases of racial classification—is one we've already discussed in detail in this book. The dispute focuses on the claim that racial categories represent genetically distinct groups as opposed to the idea that racial categorization is more of a social construct than a biological one. As Sternberg puts it:

> Where does race fit into the genetic pattern we have been discussing above? . . . In fact, it does not fit at all. Race is a socially constructed concept, not a biological one. It is a result of people's desire to classify. People seem to be natural classifiers: they try to find order in the natural world. . . . Any set of observations of course can be categorized in multiple ways. People impose categorization and classification schemes that make sense to them and, in some cases, that favor their particular, often nonscientific, goals.[28]

Classification of people into a few racial categories (usually based on self-identification) grossly oversimplifies and obscures the genetic complexities of ancestry that underlie the realities of human diversity. Therefore, claims that IQ differences between races are largely genetic is scientifically flawed from the outset.

The second issue—measuring intelligence—has generated volumes of discussion, along with a variety of hypotheses to explain the meaning of metrics used to quantify intelligence. Much of the current psychological literature is focused on measuring IQ and an associated value called g, which is proposed as a numeric representation of general intelligence.[29] Some have argued that g is a real and measurable aspect of human nature and that IQ tests reliably quantify it, particularly if components of IQ tests are weighted toward certain aspects of intelligence most relevant to g, a procedure called g-loading. Others counter that IQ and g are too narrow to encompass the complexities of human intelligence, and a one-dimensional measurement misrepresents true intelligence.

According to several reviews of the topic, there are three prominent theories. The first, often called CHC theory—after psychology professors Raymond Cattell, John Horn, and John Carroll—is a synthesis of several related theories. It posits g as a measure of general intelligence and proposes subdivision of g into g-f (fluid ability) and g-c (crystallized ability). Fluid ability is the capacity to think rapidly and deal successfully with new situations and previously unknown factors. Crystallized ability consists of the store of knowledge relevant to daily tasks that a person retains and can recall, such as vocabulary.

Modern IQ testing and measurement of g are largely based on CHC theory, measuring aspects of both g-f and g-c, and such testing serves as the foundation for a large body of research on measuring intelligence.

An alternative is Gardner's theory of multiple intelligences, named after Harvard professor Howard Gardner. He disputes the validity of g as a general measure of intelligence and instead has proposed that intelligence falls into multiple categories: linguistic, mathematical, spatial, musical, bodily-kines-thetic, interpersonal, and intrapersonal. Although Gardner's theory of mul-tiple intelligences has its proponents, a majority of psychologists have opted instead for the more simplistic CHC theory.

The third theory is the triarchic theory developed by Sternberg during the time he was a professor at Yale. It proposes three broad categories of intel-ligence: creative, analytical, and practical. He maintains that intelligence in each of these categories can be measured and that measurement of all three areas can "improve prediction of both academic and nonacademic performance in university settings and reduce ethnic-group differences." Furthermore, he argues that this theory is important for education because "teaching that incorporates the various aspects of intelligence increases academic perfor-mance relative to conventional teaching."[30]

Sternberg, along with Yale colleagues Elena Grigorenko and Kenneth Kidd, determines that "intelligence is, at this time, ill defined. Although many investigators study 'IQ' or 'g' as an operational definition of intelligence, these operationalizations are at best incomplete, even according to those who accept the constructs as useful."[31]

Critics of intelligence tests also often question the validity of IQ tests because of culture bias. Certain cognitive skills may carry a higher value in one cultural tradition than in another, and tests—however well designed to avoid cultural bias—may still favor or disfavor people on the basis of their cultural background.

Nisbett and his colleagues, in their 2012 review of the current state of psychological research on human intelligence, sum up the situation as follows:

> The measurement of intelligence is one of psychology's greatest achieve-
> ments and one of its most controversial. Critics complain that no single

test can capture the complexity of human intelligence, all measurement is imperfect, no single measure is completely free from cultural bias, and there is the potential for misuse of scores on tests of intelligence. There is some merit to all these criticisms. But we would counter that the measurement of intelligence—which has been done primarily by IQ tests—has utilitarian value because it is a reasonably good predictor of grades at school, performance at work, and many other aspects of success in life.[32]

They are quick to point out, however, that "types of intelligence other than the analytic kind examined by IQ tests certainly have a reality" and that "measuring nonanalytic aspects of intelligence could significantly improve the predictive power of intelligence tests."[33]

In spite of the inability of IQ tests to fully represent the complexity of human intelligence, and largely for their predictive value regarding academic and employment success, IQ scores are widely used as the standard for much of the research on intelligence. However, because IQ scores do not fully represent human intelligence, Sternberg has cautioned that the language used in reference to intelligence is important: IQ represents a subset of cognition and should be referred to specifically as *IQ*, not as overall intelligence.[34]

We now move to the third issue at the core of the hereditarian controversy, which is perhaps the most contentious of the three. It is sometimes called the "nature versus nurture" debate, although a more accurate term is the *heritability of intelligence*. The term *heritability* means the proportion of overall variation in a population attributable to genetic variation. Body height, for instance, varies considerably among adults. Part of this variation is attributable to genetics (the combination of variants in DNA inherited by each person that influence body growth), and part is due to environmental variation (such as poor nutrition or diseases that may stunt growth during childhood). The heritability of body height in any population is simply a numerical value that defines what proportion of the overall variation for height is attributable to underlying genetic variation. It is often expressed as a percentage value between 0 and 100. A value of 80 percent heritability for body height in a particular population, for example, implies that 80 percent of the variation is attributable to genetic variation and 20 percent to nongenetic variation, presumably environmental variation.

Few scientists dispute that both genetic and environmental effects interact to influence variation for most characteristics, including human intelligence (in whatever way it is measured). What is disputed is whether differences between racially defined groups—such as the fifteen-point difference for average IQ scores between "American blacks and whites," as stated by Herrnstein and Murray—is influenced by innate genetic differences between these groups.

As simple as it seems to be, heritability is unfortunately one of biology's most misunderstood and misapplied concepts. And most of this misunderstanding and misapplication boils down to three main points: First, heritability is a measure of *variation*, not magnitude. It tells us nothing about the degree of a trait, only what proportion of its variation can be attributed to genetic variation. Second, because it deals entirely with variation among individuals, it never applies to any single individual. Rather, it applies to each defined population in which it is measured. Third, heritability is not a permanently fixed value that can be measured in one population, then applied to another. It may vary among populations, and even over time in the same population, because environmental variation may change over time. Thus, strictly speaking, any measurement of heritability applies only to the population in which it is measured and at the time it is measured.

An often-used analogy, presented in various iterations by different authors, illustrates how heritability may be misapplied when attempting to explain racial-group differences in IQ. Perhaps the most cited version is by Harvard geneticist Richard Lewontin in a 1970 article titled "Race and Intelligence,"[35] which was a critical response to Jensen's 1969 article "How Much Can We Boost IQ and School Achievement?" Lewontin asks us to imagine taking two handfuls of seeds from the same bag of genetically diverse seed corn, such that the genetic diversity of seeds in each handful is equivalent. One handful is planted and raised in a highly uniform environment with a fertilizer solution containing all the mineral nutrients the corn plants need for optimal growth (like the liquid fertilizers for houseplants sold in department and home improvement stores). The other is planted and grown in exactly the same environment but with a suboptimal fertilizer solution that has only half the nitrogen (a major nutrient needed by plants in relatively large amounts), and half the zinc (a minor nutrient needed in trace amounts). Over time, the corn plants grown in the environment with the optimal nutrient solution vary in height due solely to their genetic differences. The same is true for the plants grown with the solution that lacks optimal amounts of nitrogen and zinc. But the *average* height of plants grown in the optimal environment is greater than the average height of plants grown in the suboptimal environment because of the difference in nutrient supply. In both cases, heritability is 100 percent for each group because all variation *within* each group is due to genetic differences. The difference *between* the averages of the two environments, however, is entirely environmental because the genetic diversity of plants in the two environments is the same. Lewontin's point is that high heritability *within* a group provides no evidence whatsoever that differences *between* groups are genetic.

He then further extends the analogy, asking us to presume that the lack of nutrients in the suboptimal environment is due to a careless laboratory worker who failed to add the proper amounts of nitrogen and zinc. A chemist is called in to diagnose the cause of the difference between the two groups and determines that one solution lacks sufficient nitrogen. The nitrogen is added, the experiment is repeated, and the difference is narrowed but not eliminated because the chemist did not test for differences in the trace quantities of zinc required for optimal growth. Lewontin's point here is that identifying some environmental causes and correcting them may not completely erase group differences until *all* environmental causes have been identified and corrected. Thus, failure to erase a gap through environmental intervention does not necessarily offer evidence of a genetic cause for that gap.

Lewontin's analogy with corn plants is a good visual representation of heritability, and it also represents the concept's origin. Heritability and methods for measuring it arose from plant and animal breeding, where its utility is very powerful. When heritability is high, plant and animal breeders can rapidly improve inherited characteristics through artificial selection because they are able to select individuals that are genetically most predisposed to inherit genes conferring the traits they wish to improve. When heritability is low, however, progress from artificial selection is slow and unproductive because environmental differences mask genetic differences. Therefore, much of modern plant and animal breeding is aimed at implementing measures to maximize heritability, usually through reducing environmental variation, such as planting experimental plants in environments that are as uniform as possible or providing experimental animals with adequate food in exactly the same rations for each individual. Plant- and animal-breeding experiments designed to measure heritability are usually highly controlled, with elaborate statistical and experimental designs to ensure accurate results.

By contrast, measuring heritability in humans is notoriously challenging. Researchers, for obvious ethical reasons, must not subject humans to highly controlled environments as they do for experimental plants and animals. And even if they were to do so, the results would have little meaning because heritability in the real world is dependent on actual environments, not experimental conditions. As Nisbett and his colleagues put it in their 2012 review:

> The concept of heritability has its origins in animal [and plant] breeding, where variation in the genotype and environment is under the control of the experimenter, and under these conditions the concept has some real-world applications. In free-ranging humans, however, variability is uncontrolled, there is no "true" degree of variation to estimate, and heritability can take practically any value for any trait depending on the relative variability of genetic endowment and environment in the population being studied.[36]

Some of the most common estimates of heritability in humans are based on studies of identical twins who were separated at birth and raised apart. The idea here is that identical twins are *genetically* identical. Therefore, any differences between them are assumed to be entirely environmental (in other words, heritability is zero for identical-twin pairs). Measurement of how much two twins differ for any characteristic, therefore, offers an estimate of purely environmental variation between them. By combining these differences among multiple pairs of identical twins reared apart, researchers can derive an average estimate of environmental variation for the environments of these twin pairs without the confounding effects of genetic variation. If they then apply these estimates of environmental variations to people who *do* differ genetically but are raised in similar environments to the twins, they can derive estimates of heritability for the people who are genetically different. The assumption— which is key—is that any variation greater than that observed between twins must be genetic. In some cases, identical twin pairs are compared to fraternal twin pairs, who differ genetically as much as full siblings but have the comparative advantage of having shared the same womb and been born on the same day, like identical twins. Alternatively, researchers can compare variation among relatives with different degrees of genetic similarity—such as full siblings, half siblings, and cousins—who are raised in similar environments as a means of statistically estimating heritability.

Regarding such estimates of heritability, Herrnstein and Murray conclude that

> the genetic component of IQ is unlikely to be smaller than 40 percent or higher than 80 percent. The most unambiguous direct estimates, based on identical twins raised apart, produce some of the highest estimates of heritability. For purposes of this discussion [differences between racial groups], we will adopt a middling estimate of 60 percent heritability, which, by extension, means that IQ is about 40 percent a matter of environment.[37]

Critics of *The Bell Curve* rightly point to this generalization of "a middling estimate of 60 percent heritability" as a gross misapplication of the concept. According to Nisbett and colleagues' 2012 review of current research, "the heritability of intelligence test scores is apparently not constant across different races or socioeconomic classes,"[38] reinforcing the long-standing caveat that heritability should not be generalized from one population to another.

Moreover, reliance on estimates of heritability based on environmental variation measured in identical twins has also been criticized because, even when twins are reared apart, environmental similarities may be confounded with the genetic identity of twins. For example, adopted children are typically raised in more affluent homes in environments that tend to be more uniform and less representative of overall environmental variation, thus resulting in potential overestimates of heritability on the basis of twin studies.

Also, the shared uterine environment of twins (identical or fraternal), which is unrepresentative of people who are not twins, can have a confounding effect on heritability estimates, even when identical and fraternal twins are compared. Identical twins typically share the same placenta, whereas fraternal twins have separate placentas, resulting in different uterine environments.

The uterine environment is especially pertinent to measurements of IQ later in life, particularly if prenatal care is poor or if a birth mother has abused alcohol, tobacco, or illicit drugs during pregnancy because these factors can permanently affect brain development, and, in the case of twins, such factors may or may not affect the twins similarly, leading to misinterpretation about the degree of genetic influence on the brain.

Similar obstacles confront researchers who conduct nontwin studies for measuring heritability in humans. One of the most serious obstacles has been the inability to directly estimate genetic variation. Historically, researchers have estimated genetic variation by the degree of relatedness of subjects, such as comparing identical twins, full siblings (including fraternal twins), half siblings, and cousins to people who are unrelated. Such indirect measures of genetic variation are highly problematic, however. For instance, the degree of genetic variation between full siblings is not generalizable because it depends on how genetically different their lines of ancestry are. For instance, it is often said, as Herrnstein and Murray put it, that "full siblings share about 50 percent of genes."[39] Such a statement represents a seriously naïve understanding of human genetics. In reality, full siblings share about 50 percent *of the variants that are heterozygous in each parent* and 100 percent of the variants that are homozygous in each parent. The more diverse the lines of ancestry are for an individual's parents, the greater the degree of heterozygosity in the parents and the greater the degree of genetic variation in their children. Therefore, genetic variation among full siblings varies considerably from one family to another; full siblings whose parents have similar ancestry are genetically more similar to one another than full siblings whose parents have genetically diverse ancestry. The same holds true for half siblings, cousins, and other people who are genetically related. Differing degrees of genetic diversity from one family to another result in different heritabilities among biological full siblings or other relatives, further dispelling the notion that heritabilities in humans can be generalized.

To their credit, many scientists who conduct research on heritability in humans recognize these and other limitations and clarify them when reporting research. Scientists often use complex and elaborate scientific and statistical methods to derive estimates of heritability that are as reliable as possible despite the large margins of error that are typical of heritability measures in humans. Most of these scientists are well-trained researchers who appropriately apply scientific methods to their work. Unfortunately, not everyone is as attentive; overly simplistic generalizations of heritability for IQ in humans often lead those who advocate hereditarian models to unsubstantiated conclusions that may be based more on political ideology than on scientific evidence.

Published studies on IQ and the interaction between genetic and environmental variation often report contrary conclusions. For instance, Herrnstein and Murray determined that, according to the studies they reviewed, socioeconomic status and shared environment (the fact that children raised in the same family share many of the same environmental influences) had little effect on heritability for IQ. Rushton and Jensen arrived at similar conclusions. By contrast, other researchers offer evidence that socioeconomic status and shared environment have a considerable effect on heritability for IQ. These latter studies often, though not always, portray a positive correlation between heritability and socioeconomic status, especially when shared environment is taken into account. Heritabilities tend to be highest in families in which parents are well educated or families of high socioeconomic status, whereas heritabilities are lowest for impoverished families and when parents have attained little education.[40] However, yet other studies have led researchers to arrive at the opposite conclusion: heritabilities are highest in families of low socioeconomic status when compared to the more affluent.[41]

In many of these contradictory studies, there is nothing wrong with the science or the interpretations. In fact, it is no surprise that studies on heritability of IQ may show widely differing results. They often are conducted on different populations in different places with people whose degree of genetic relatedness differs, and with subjects of different age groups and educational backgrounds. They are excellent examples of the malleability of heritability estimates in humans. By focusing on a selected subset of studies, advocates of a particular point of view can assemble what seems to be ample evidence to support their preferred models, when, in fact, the real situation is highly complex, varied, and inconclusive.

The most reliable compilations of data are those that identify major trends among exceptionally large numbers of people over decades of time. One such trend is known as the *Flynn effect*, named after psychologist James R. Flynn of the University of Otago in New Zealand. His own words best describe it: "'The Flynn Effect' is the name that has become attached to an exciting development, namely, that the twentieth century saw massive IQ gains from one generation to another."[42]

These gains are evident in essentially every country where IQ tests have been administered long enough to detect a trend—thirty countries as of 2012.[43] In countries that had been modernized by the beginning of the twentieth century, the gains have averaged approximately three IQ points per decade.[44] Less developed countries have also almost universally experienced gains.[45] These gains, however, appear to be peaking in the most highly developed countries, such as Sweden. According to the 2012 review article by Nisbett and colleagues (one of whom is Flynn), "If Sweden represents the asymptote that we are likely to see for modern nations' gains, the IQ gap between developing and developed nations could close by the end of the 21st century and falsify the hypothesis that some nations lack the intelligence to fully industrialize."[46]

As psychologists and geneticists have pointed out, genetic changes over so short a period are insufficient to explain the Flynn effect, and, therefore, changing environments must be its principal cause. A number of environmental causes have been proposed, such as improvements in nutrition and education. Notably, the gains are not equivalent across the board but are, in most cases, greatest among groups of people who historically have scored lowest on IQ tests, and scores dramatically improve as economic and educational status improves. According to Nisbett and colleagues, "It seems likely that the ultimate cause of IQ gains is the Industrial Revolution, which produced a need for increased intellectual skills that modern societies somehow rose to meet."[47] Flynn, as a coauthor, agreed, but he also proposes a more specific explanation consistent with the Industrial Revolution and changes in education. He notes that gains are not evenly distributed across the subject areas that IQ tests measure but, instead, are greatest for analytical reasoning and pattern recognition and lowest for basic arithmetic. Education during the latter part of the twentieth century has placed greater emphasis on logic and abstract reasoning, mirroring the areas of IQ for which nations experienced the greatest gains.[48]

Perhaps the most serious deficiency for arguments supporting hereditarianism, especially racial hereditarianism, is the fact that nearly all measurements of genetic variation for IQ have been indirect. Genetic variation has been assigned essentially by default, under the presumption that any variation unexplained by environment must be genetic. One of the greatest advances in human genetics in recent years is the ability to measure genetic variation *directly* by tracking variants in DNA. In previous chapters, we've already seen how direct detection of variants in DNA can explain genetic variation for skin, hair, and eye pigmentation; inherited conditions; and susceptibility to a wide range of diseases, both infectious and noninfectious.

The ultimate confirmation of the underlying genetic influence on variation for any human trait is identification of causal variants in DNA. In the late 1990s, a number of psychologists expressed hope that genes and variants governing variation for intelligence would soon be identified.[49] However, in spite of much research, identification of such genes and variants has proven elusive. Robert Plomin of King's College London has contributed some of the most significant research on variation for IQ, particularly through large-scale and long-term twin studies in the United Kingdom. In a 2013 article, Plomin laments that predictions made in the late 1990s regarding gene identification were "overly optimistic."[50] He also refers to the problem of *missing heritability*, defined as failure to confirm heritability estimates with DNA analysis:

> Genetic designs such as the twin method would no longer be needed if it were possible to identify all of the genes responsible for heritability. However, it has proven more difficult than expected to identify genes for complex traits, including g, which has led to the refrain of "missing heritability."[51]

During the first decade of the twenty-first century, a few genes and variants were identified as being associated with variation for intelligence. However, in 2011, a group of sixteen scientists representing institutions in the United States and Europe published a large-scale study that attempted to confirm these associations. They phrased the title of their article as a sentence, which aptly summarizes their main conclusion: "Most reported genetic associations with general intelligence are probably false positives."[52]

Thus far, no single variant in any gene has been definitively associated with variation for IQ or *g*. However, when multiple variants are examined in the aggregate, a statistically detectable association with IQ and *g* has been identified. However, each variant appears to have no more than a very minor effect on variation for IQ or *g*. Only when many variants are examined in large numbers of individuals in the aggregate can a genetic influence on variation for IQ and *g* be detected.

Although large-scale studies on DNA variants and their relationship to IQ and *g* are relatively recent and few, a consensus on two major points appears to be emerging. First, variation for IQ and *g* is heritable to some degree, with heritability estimates differing among the populations studied.[53] In this sense, direct estimates of genetic variation based on DNA studies are consistent with earlier studies relying on indirect estimates, which likewise produce varying heritability estimates. Second, no single gene or variant appears to have a major effect on variation for IQ or *g*. Instead, a large number of DNA variants, each with a very minor effect, combine in complex ways to influence variation for IQ and *g*. Adding to this complexity is the significant and often-changing influence of environmental variation.

That variation for intelligence is probably heritable to some degree has never been in question, even from the staunchest opponents of hereditarianism. Gould, for instance, wrote in *The Mismeasure of Man*,

> The hereditarian fallacy is not the simple claim that IQ is to some degree "heritable." I have no doubt that it is, though the degree has clearly been exaggerated by the most avid hereditarians. It is hard to find any broad aspect of human performance or anatomy that has no heritable component at all.[54]

Instead, the principal, and most controversial, question is whether average differences in IQ between so-called racial or ethnic groups can be attributed to genetic differences between these groups. DNA studies have thus far provided little clarification on this question. The vast majority of people who were the subjects of DNA-based studies were drawn from populations of predominantly European ancestry (mostly from the United Kingdom, Australia, the Netherlands, and the United States).[55] The failure to identify any variants with major effects makes it difficult to extend the results of these studies to more ancestrally diverse populations.

Much of human genetic variation consists of ancient African variation. The inference that genetic variants influencing human intelligence are numerous, each with a very small effect, further suggests that most of these variants are dispersed throughout humanity rather than concentrated in people with a particular ancestral background. The fact that DNA studies on IQ and *g* have yet to include large numbers of people with highly varied ancestry means there currently is no direct DNA evidence supporting the claim that average differences between ethnically defined groups have a genetic basis. And, given what *is* known about the predominantly ancient African nature of the majority of human genetic diversity, such evidence is unlikely to be discovered. Concurrently, there is abundant evidence that a host of nongenetic factors influence variation for IQ and *g*. The Flynn effect, in particular, provides definitive historical evidence that large gains are possible when overall economic and educational conditions improve.

Current scientific evidence fails to support hereditarian arguments against public investment in education and programs that benefit the economically disadvantaged. There is ample evidence that investment in education yields economic benefits that derive from a well-educated workforce. And there is no question that disparities for educational and economic opportunities have conferred substantial disadvantages to people in ethnic minorities—in the United States, especially African, Hispanic, and Native Americans—a fact readily acknowledged by even the most devout hereditarians.[56]

Yet, as I write this, disinvestment in public education has been underway for some time, from the preschool through higher-education levels. Head Start, a long-standing program in the United States to assist disadvantaged preschool children, suffered one of its largest cuts in 2013, depriving thousands of impoverished children of early educational intervention. Public schools have likewise suffered reductions in numbers of teachers, budgets for facilities and supplies, and educational programs. Public colleges and universities are especially hard hit as increasingly smaller proportions of their budgets come from public funding. As a professor and administrator at an open-enrollment public university whose mission is to serve diverse groups of students, I have a firsthand sense of how seriously such disinvestment impacts student success.

These cuts cannot be attributed to the ideologies of hereditarians but, rather, the difficult choices politicians and government administrators must make when faced with competing demands for limited budgets. Nonetheless, the impact of educational disinvestment is the same regardless of the reason. Much of today's educational inequality is the legacy of historical racism. The affluence of one's family, the neighborhood in which one resides, and the educational background of one's parents all are major factors influencing educational opportunity and attainment. The historic roots of educational inequality lie in the history of racial segregation, and, inevitably, they continue to discriminate along socially defined lines of race, even if the intent to do so is a relic of the past. Recent trends of reversion to increased educational inequality prompted by budget woes disproportionally affect the disadvantaged, further exacerbate attainment gaps, and, tragically, erase gains that have started to close those gaps.

CHAPTER 7
THE PERCEPTION OF RACE

Not long ago, I had the privilege of traveling to northern Peru with a colleague who is a biological anthropologist. Our destination was the city of Lambayeque, where, at a nearby archaeological site, the remains of an elite woman who lived eight hundred years ago had been discovered and were being excavated. She was buried in a deep and elaborately decorated tomb near a colossal adobe pyramid. Her body had been draped in a blanket with dozens of copper medallions sewn into it. She wore layered pectoral necklaces on her chest: one made of copper ornaments, one of carved shell plates, another containing thousands of tiny colored beads meticulously fashioned from shell that had been strung together. Her arms bore bracelets of gold, and in one hand she held a gold scepter. A copper mask covered her face, and on her head was an ornate crown. Though her ears had long since decayed, it was apparent that the lobes had been gauged and stretched to hold exquisitely tooled gold earspools.

By then, all soft body tissue and fabric had deteriorated. All that remained were her teeth, the fragmented bones of her body, and the shell and metallic pieces of her ornamentation. On the forehead of her skull were traces of a bright vermilion pigment called cinnabar, a brilliantly colored but toxic mercury compound that had been smeared onto her face before burial. Accompanying her were six other skeletons: her consorts who had committed ritual suicide to join her in the afterlife. We do not know the name she had during her life; today, she is known as the *Sacerdotisa de Chornancap* (the Priestess of Chornancap).

At the time of our arrival, the rich ornamentation covering her body had been painstakingly removed and catalogued, exposing her fragmented skull (figure 7.1). I had the honor of reconstructing her facial features from the skull fragments using forensic-sculpture methods. With the assistance of a student, I precisely replicated each fragment in three dimensions. Back in my studio in the United States, I assembled the replicated pieces and filled in missing sections with clay to reconstruct the skull. I placed prosthetic eyes in her eye sockets, then used colored clay to sculpt the muscles, tendons, ligaments, glands, fat, and skin. We then returned to Peru to complete the reconstruction with Peruvian colleagues from the museum that houses her remains. We gave her a wig of real human hair styled with bangs and braids, as women were depicted in the sculpted faces on pottery jars made during the time she lived. We dressed her with the ancient carved-shell beads of a pectoral necklace that had been restrung, along with her actual gold earspools (figure 7.2). A video of the process is at https://www.youtube.com/watch?v=KxFBRFXxlmQ.

Figure 7.1. Fragmented skull of an elite woman who governed the northern coastal region of what is now Peru approximately eight hundred years ago. *Photograph by Haagen Klaus; used with permission.*

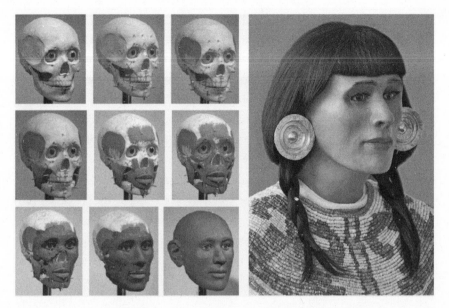

Figure 7.2. Forensic facial reconstruction of the Priestess of Chornancap, on display in the Hans Brüning National Archaeological Museum in Lambayeque, Peru. *Photographs by Daniel J. Fairbanks and Haagen Klaus; used with permission.*

As I held the fragments of her skull in my gloved hands, I thought of my ancient kinship with her. Our shared ancestry was very distant, dating back at least thirty thousand years to people who lived probably somewhere near the Caucasus Mountains in west-central Asia. Our ancient ancestors diverged at that place long ago, hers migrating eastward, mine westward. People in her ancestral lineage traversed the continent of Asia over hundreds of generations to the ancient Isthmus of Beringia—connecting what are now northeast Russia and Alaska—and crossed into North America. Their descendants continued migrating southward through North America, some of them living probably no more than a day's drive of where I now live. Eventually, her still-distant ancestors reached the western shores of South America just south of the equator. Generations later, she was born and raised to be an elite ruler adored by her people. By then, my distant ancestors had migrated into Europe, where their descendants remained for millennia. When she was alive, all my ancestors were in northern Europe, many of them in England about the time when the Magna Carta was written and sealed. They were entirely unaware the continent where she lived even existed. Eight centuries later, I was in Peru holding her remains, bringing her image back to life—my distant cousin separated by hundreds of years and thousands of generations.

Piecing together scientific information from ancient remains, historic accounts recorded during the Spanish conquest, and the native languages of modern people in the region, we now know much about her people and the civilizations that occupied that part of South America. The rich evidence reveals fascinating histories of people who migrated in and out of the region, prospered from agriculture and the sea, conducted mutually beneficial economic trade with other cultures, suffered from disease and climatic disasters, battled in war, and conquered or were conquered throughout millennia. The same can be said for ancient people in other parts of the world where abundant evidence has provided us with much information about their histories.

Today, we can add DNA analysis to the wealth of evidence from other sources to reconstruct the history of humanity. Most of the information from DNA comes from the modern descendants of people who lived long ago, although, in a few cases, DNA extracted from the remains of ancient people themselves has provided direct evidence of their genetic constitution. Moreover, DNA from the plant and animal species they domesticated and biogeographical information revealing how those species spread throughout the world offer additional evidence of ancient human history, migrations, and settlement. We are now at a point where we can combine information from Earth's climatic history, archaeology, anthropology, paleontology, plant and animal domestication, linguistics, and large-scale analysis of DNA to determine how humans populated the world and reconstruct the complexities of their ancient migrations. This history reveals how the ancestors of today's worldwide human population transitioned from hunter-gatherers to agriculturalists; how vast civilizations rose and fell; how some conquered others, taking over large regions; how factors such as climate change and infectious disease decimated certain populations, yet spared others; and how humans have moved (voluntarily or involuntarily) across and between continents, ultimately resulting in today's largest, most widespread, most mobile, most diverse, and most genetically complex population in human history.

An enormous amount of recent information became available with the development of laboratory methods that allow geneticists to examine hundreds of thousands of DNA variants simultaneously, generating massive data sets from which they reconstruct these histories. As a geneticist who has worked with these methods since the 1980s, I have experienced this astonishing progress firsthand.

Although the methods for analyzing these data sets are complex, the underlying premise is simple: All DNA variants originate as mutations that can happen in anyone at any time in any place. When a new variant originates, it is superimposed on a particular genetic background of nearby variants, and comparison of any variant with its background of nearby variants in DNA from people who are native to particular regions can help identify approximately when and where a particular variant originated.

The most advanced and comprehensive worldwide study of DNA variants to date was published in February 2014 by a group of geneticists and statisticians headed by Simon Myers of Oxford University. They examined 474,491 variants in 1,490 people from ninety-five populations sampled from around the world. Their analysis determined which human populations had contributed particular variants to other populations and approximately when those contributions happened on a worldwide scale. If one population contributes variants to another, the descendants in the recipient population are said to be *genetically admixed*, hence the title of their study: "A Genetic Atlas of Human Admixture History."[1]

Both archaeology and DNA histories reveal that for most of human history, population sizes were relatively small, and migration events took hundreds to thousands of years, spanning many generations. Tens of thousands of years ago, human populations were slowly expanding into previously unoccupied parts of the world. Their weapons were crude, transportation was mostly by foot, and sustenance was from hunting and gathering. Skirmishes were relatively small and localized, without the wars and massive conquests typical of more recent human history. As new DNA variants arose, they accumulated in localized populations over many generations. Although older DNA variants had traveled far over tens of thousands of years and hundreds of generations, new DNA variants tended to remain localized to the regions where they originated, accumulating along with other variants against identifiable genetic backgrounds. To spread far, they required long periods of time spanning numerous generations. These newer variants are the ancestry informative markers we've discussed in earlier chapters, each of which points to a particular part of the world as its ancestral source. Although most of human genetic variation is ancient African and is dispersed throughout humanity, by focusing on these more recent ancestry informative variants, scientists can decipher approximately when and where major human migration events took place.

According to these DNA studies, and congruent with recorded history, the localized distributions of these relatively recent variants began to change about four thousand years ago. Human mobility dramatically increased as the world's population size expanded, advanced civilizations emerged, long-range transportation technologies were invented, economic trade grew to an intercontinental scale, and large armies conquered people residing across vast regions. Newer variants that had previously been localized began spreading through major migration events, some dispersed to parts of the world very distant from their origins. As people who entered a distant region mated with those who were there, or took people from those regions back to their lands of origin, DNA variants became intermingled in their offspring and were permanently introduced into the offspring of people in different parts of the world.

"A Genetic Atlas of Human Admixture History," together with several earlier studies focused on particular geographic regions, has uncovered strong evidence of more than a hundred of these genetic dispersal events, affecting the vast majority of the world's population, during the past four thousand years. One of the most significant is the Bantu expansion in sub-Saharan Africa, which intermingled variants carried by the migrating Bantu agriculturalists with the variants present in the highly diverse native populations of the African subcontinent. Another is the Mongol Empire, initiated by the conquests of Genghis Khan. At its peak in the thirteenth century, it spread from Asia's eastern shores to what is now Turkey in the west. Today, DNA variants carried by Mongol invaders remain intermingled with the variants that originated in these regions, along with ancient African variation. The Arab slave trade from the seventh through the nineteenth centuries brought slaves captured in Europe and Africa to the Middle East. The spread of variants went both ways: slave traders who lived in Europe and Africa left descendants, and descendants of slaves taken to the Middle East were eventually integrated into the populations there. As a consequence, people from much of the Middle East carry African variants. The Arab conquest of north Africa began shortly after the death of Mohammed in 632 CE. Muslim armies consisting mostly of people from the Arabian Peninsula—but also converts to Islam from more distant regions, including Romans and Greeks—conquered Egypt and eventually spread into lands southward along the Nile River and westward along the Mediterranean coastline.[2] The genetic constitution of people throughout north Africa strongly bears the mark of this conquest. The Greek conquest of Alexander the Great during the fourth century BCE extended from Greece into Egypt in the south to what is now Pakistan in the east. The Roman Empire, at its peak in the second century CE, spread over nearly all of Europe and into the Middle East and north Africa, with the resulting dispersion of variants from Roman invaders throughout that empire. The Ottoman Empire, ruled from Constantinople (modern-day Istanbul), occupied much of what had been the eastern and southern Roman Empire, reaching its peak in the seventeenth century CE. And, last, European colonialism and the Atlantic slave trade were the most significant genetic dispersal events in modern times, spreading variants carried by European and African immigrants into distant parts of the world. The overwhelming majority of people on Earth have inherited a mix of variants tracing to major migration events such as these.

Other migration events apparent in the DNA evidence do not coincide with any known historical events but nonetheless happened, their only existing record consisting of recently discovered evidence in DNA. You can access the information from this research in an easy-to-use interactive map showing the worldwide dispersal patterns of DNA at http://admixturemap. paintmychromosomes.com.[3]

Each major historic migration event reshuffled genetic variants, reshaping the genetic structure of populations across and between continents all over the world. Each ancestry informative variant typically remained most prevalent in its region of origin, but many of these variants were dispersed elsewhere, found today in people across wide geographic areas.

There is also ample evidence from history that major migrations resulted in racial subjugation. In most cases, those migrating into a region, often as members of an invading military power or as colonists, considered themselves genetically superior to those who were already there, claiming an inherent right to rule over those they subjugated. Over time, as civilizations rose and fell and generations passed on, the DNA variants brought by invaders or immigrants intermingled with the variants that had been present in the people who were already there, the resulting admixture persisting for generations to their descendants alive today.

European colonialism and the Atlantic slave trade, reaching their peak in the seventeenth and eighteenth centuries, were among the most overwhelming genetic dispersion events humanity has ever experienced. Many of you reading this book trace at least some of your ancestry to people who either chose or were forced to leave their native lands to live in distant parts of the world as a consequence of these events. I count myself in this group. All of my ancestral lines trace to people who left the British Isles, Scandinavia, and the northern coast of continental Europe at one time or another for the shores of North America.

To many, the idea that there are no distinct biological races seems counterintuitive. In the United States, for instance, it is not uncommon for people to self-classify as white, black, Asian, Native American, Pacific Islander, or any other number of racial categories and to have genetic testing with ancestry informative markers confirm those classifications, at least in part. Does this not imply that some sort of racial boundaries exist, perhaps major continental boundaries, such as European, Asian, African, and Native American? In fact, we're about to see why the perception of distinct racial categories is more an artifact of immigration history than a consequence of true biological boundaries. To see how this happened, let's visit three places where the history of modern racism is well known—South Africa, Australia, and the United States—and follow the histories of people now living in these places, from prehistoric times to the present. We'll start with South Africa.

The first modern humans to settle the southernmost regions of Africa were the ancestors of the San and the Khoikhoi people, sometimes referred to collectively as the Khoisan ("Khoi" and "San" combined). According to DNA analysis, the ancient ancestors of these people diverged from the ancestors of all other people as early as 140,000 years ago.[4] The oldest archaeological remains discovered thus far in the southernmost parts of Africa date to about

44,000 years ago and probably belong to the ancient ancestors of the San people (the Khoikhoi arrived in the region more recently). The native languages of the Khoisan constitute a group of African click languages, which have a clicking consonant sound produced by sucking in air through closed lips. The click is sometimes written as !, as in the !Kung language spoken by the !Kung people of the Kalahari.

The first Europeans to land on the southern tip of Africa were on a Portuguese ship that reached what is now the Cape of Good Hope, near the southern tip of Africa, in 1488. This opened the door to what eventually became a major seafaring trade route from Europe to India around the cape. As the economic importance of this trade route grew over the next century and a half, the Dutch East India Company, headquartered in Amsterdam, came to dominate this trade. It established a resupply post for ships at the cape in 1652 with a small group of Dutch settlers. Some of the settlers moved into the surrounding lands to establish farms that could supply food and other goods for the passing ships. Over time, the settlement grew, eventually becoming the city of Cape Town, and Dutch farms were established around it. Eventually, German, Scandinavian, and French Huguenot immigrants joined the Dutch settlers, creating colonies of northern Europeans.

The first native people the European colonists encountered were the Khoikhoi, whom they named "Hottentots," and the San, whom they named "Bushmen" (the terms in quotation marks are now considered pejorative and offensive). The European colonists and the native Khoisan people were from two geographic extremes, northern Europe and southern Africa. Although both groups were unaware of it, they shared common ancestors who, more than 140,000 years earlier, belonged to the same ancient population of people living in Africa.

Though separated for thousands of generations, and very different culturally, the European immigrants and Khoisan people were not so different genetically, sharing numerous variants dating back to their common ancestry. For instance, among the Europeans, some people had type A and others type O blood, and the same was true of the Khoisan, with type O the most common in both groups.[5] Most of the variation in both groups was likewise ancient African and was shared. However, since the time of ancestral separation, newer genetic variants had accumulated independently in both groups, among them variants conferring differences in skin, hair, and eye pigmentation; hair texture; facial features; and body structure. For instance, the Dutch were likely then among the tallest people in the world and the Khoisan among the shortest, as is still the case.

Although the Khoisan people and European immigrants differed genetically, they also differed for nongenetic characteristics, such as culture, clothing, language, religion, technology, weapons, and how they obtained food—differences that, like the genetic differences, had accumulated over countless generations of geographic separation. The two groups were discontinuous and distinct, for culture and for a minority of their genes. They perceived themselves as separate and distinct races.

However, going back to the time before European colonization, if we were to trace human genetic diversity along a land route from the Khoisan in southern Africa to the Dutch in northern Europe (passing northward through Africa, across the Sinai into the Middle East, then through what was then the Turkish Ottoman Empire and the Balkans into northern Europe), a complex pattern of changing human genetic diversity would have been evident. The transitions would have been gradual, with evidence of admixture and variation along the way, but without boundaries delineating distinct genetic races of people. The reason for the perception of distinct races when European colonists settled southern Africa was the sudden juxtaposition of people from the extremes of that long continuum.

European merchant ships returning from India to Holland stopped at the Cape Town outpost for resupply. Among the "cargo" these ships carried were people taken as slaves, mostly from India. As some of these slaves were traded or sold to the settlers, another distinct group of people—from south Asia—expanded in this part of Africa. They were physically, culturally, and genetically distinct from the European settlers and the native Khoisan people and were perceived as another separate and distinct race.

The consequences of these differences were devastating. Consistent with their worldview and religious convictions, the European settlers saw themselves as intellectually, culturally, and genetically superior, with a divinely appointed destiny. Though they were fewer in number than the Khoisan peoples, their weaponry was far more powerful, and they quickly overcame any resistance. More devastating than weapons, however, were the strains of infectious diseases the Europeans brought, to which they were more resistant than the Khoisan were because ongoing exposure had stimulated their immune systems. One of the most tragic consequences of early European immigration into southern Africa was the number of people who died of infectious disease carried by the Europeans, especially smallpox. In 1713, smallpox-causing viruses from the contaminated laundry of a passing ship first infected the south Asian slaves working in the laundry and, shortly thereafter, infected the European settlers. Though some from both those groups died of smallpox, the most severe outbreak was among the Khoikhoi. They had never been exposed to smallpox, so their immune systems had never had a chance to produce protective antibodies against the disease-causing virus. This epidemic, followed by two subsequent smallpox epidemics, devastated Khoikhoi populations. In the end, the Khoikhoi people and their culture were essentially annihilated in southern Africa. Jared Diamond's title for his Pulitzer Prize–winning book, *Guns, Germs, and Steel*, aptly summarizes in three words why European colonists overwhelmed native people in Africa and other continents, with germs initially the most devastating of the three.

As the number of Europeans increased in southern Africa, those who were farmers and herders began migrating northward and eastward, establishing farms and ranches. Tensions arose when the British seized control of the cape and drove the non-British European farmers and ranchers, known as Voortrekkers (meaning "forward-trekkers" or "pioneers"), even farther north and east into parts of what are now South Africa, Lesotho, Swaziland, Zimbabwe, and Mozambique. The British eventually established their own settlements in these lands as well.

The expanding northeastward settlement and land use encroached into the lands of Bantu people. The Bantu had originated tens of thousands of years earlier, far to the northwest in what are now Cameroon and Nigeria. About 3,500 years ago, they transitioned from a hunter-gatherer society to an agriculturalist society, with food production focused on yams, a nutritious and high-energy food. This transition allowed Bantu population sizes to increase well beyond what hunter-gatherer populations could sustain.

This was the beginning of the Bantu expansion, one of the most overwhelming genetic upheavals in the history of Africa. Some Bantu populations migrated eastward, over many generations, through the Sahel region, a grassland between the Sahara Desert and the rain forests of central Africa, well suited to expanding yam cultivation. Other Bantu groups migrated southward along and near the Atlantic coastline. As they migrated, the Bantu overtook the hunter-gatherer and herder-farmer societies they encountered. Whether the assimilation of other populations was peaceful or violent, or some combination of the two, is unknown. What is known is that DNA variants that originated in the Bantu became a major part of the genetic structure of native populations along the way. By about 1000 CE, the Bantu expansion along the southeast part of the African continent had reached the southern parts, driving the Khoisan people westward into more arid regions near the cape, where the Europeans first encountered them.

Among the Bantu people the Voortrekkers confronted were the Xhosa and the Zulu, both of whom were militarily powerful. A series of wars ensued, enflamed by broken treaties and massacres, resulting in deaths on both sides. One of the most famous events was the Battle of Blood River between the Voortrekkers and the Zulu. Following the execution of a Voortrekker negotiating group carried out by the Zulu leader, both sides prepared for battle. As the time approached, the Voortrekkers made a vow to God that they would build a chapel in commemoration if they were victorious. The Zulu warriors outnumbered the Voortrekkers by more than sixty to one. The Zulu spears and shields, however, were no match for the muskets and cannons of the Voortrekkers, who fought from a defensive position behind a fortified wall of wagons. The Zulu defeat was catastrophic. More than three thousand Zulu warriors died, whereas only three Voortrekkers were wounded. At the height of the battle, Zulu blood stained the river red, hence the battle's name. The Voortrekkers viewed the victory as a divine affirmation of their destiny in the land.[6]

This was one of countless events during European colonialism that fueled the conviction of divinely appointed destiny in nearly every place in the world where European colonies existed. In southern Africa, battles between Europeans and native Africans fanned the flames of intense racial hatred on both sides. Over time, a legal system of racial suppression arose, culminating during the latter half of the twentieth century in what is famously known as *apartheid*. The historic superimposition of distinct immigrant cultures—

European, south Asian, and Bantu—into a region occupied by the Khoisan led to the perception of distinct races that persisted for generations.

Apartheid officially began in 1948 and lasted until 1994. People were legally classified as belonging to one of four racial categories: White, Black, Indian, or Coloured. Descendants of European colonists were categorized as White; Bantu as Black; south Asian as Indian; and Khoisan, along with people of mixed ancestry, as Coloured. Segregation and antimiscegenation laws separated people into their respective racial categorizations, with political and economic power concentrated in the White class. Those categorized into one of the three nonwhite classes were forced to move into racially segregated communities when the areas where they lived were officially declared as White.

The period of apartheid was long and oppressive. Nelson Mandela was imprisoned for twenty-seven years for leading nonviolent opposition and was released in 1990 as political and economic pressure throughout the world was growing against apartheid. For the next several years, Mandela negotiated the end of apartheid with then-president Frederik Willem de Klerk. In 1993, Mandela received the Nobel Peace Prize, and the following year, he was elected president in the first elections to allow nonwhite voting. News of his passing reached me as I was writing this chapter.

Political and cultural change over the past quarter century in South Africa has been rapid. Nonetheless, the legacy of racial tensions that trace their foundations to the historic juxtaposition of distinct groups of people persists. For instance, income inequality in South Africa is among the highest in the world, most of it divided along the apartheid-era racial classifications, with the highest income concentrated in the white class.[7] This same sort of income stratification is evident in most nations where European colonization brought together people of distant geographic origins, one of which is where we turn next.

The southeast coast of Australia, where the city of Sydney now stands, is the first place on that continent colonized by Europeans. The people who occupied Australia before European colonization were descended from a very ancient ancestral lineage. According to DNA analysis, some of the first people to split away from the descendants of those who left Africa more than sixty thousand years ago and migrated into the Middle East were the ancestors of Australian Aborigines. They diverged from the main population, migrating eastward, between sixty-two thousand and seventy-five thousand years ago.[8] Their descendants crossed southern Asia to southeastern Asia over a period of thousands of years. Along the way, there was limited mating in central Asia between these migrants and Denisovans, a now-extinct humanlike group closely related to Neanderthals that we know only from a few bones and DNA. By then, the migrants already carried a small proportion of Neanderthal DNA from limited mating between their ancestors and Neanderthals in the Middle East. DNA variants inherited from Denisovans are now present in people who have Aboriginal Australian, Papuan, Southeast Asian, and Pacific Island ancestry.

At the time, sea levels were much lower than they are now. A large peninsula called Sunda connected what are now the islands of Sumatra, Borneo, and Java with the Asian continent as a single landmass (figure 7.3). Some of these people crossed the narrow straits of water between Sunda and the continent of Sahul (now Australia and the islands of Papua and Tasmania) as early as fifty thousand years ago.[9] As sea levels rose when the ice age ended, the current islands in the region were formed.

After sea levels rose, the descendants of ancient immigrants remained mostly isolated in Australia for tens of thousands of years. By the time European colonists reached Australia, growth of Australian Aboriginal populations had been rising, and hunter-gatherer tribes occupied much of the continent, having populated it long ago.[10] Though they had not developed cultivated agriculture or advanced weaponry, they had rich and varied artistic and musical traditions. Magnificent rock art, sculpture, and some of the world's oldest known musical instruments are the legacies of these people and their cultures.

British colonization of Australia began at Sydney in the late eighteenth century, initially as a penal colony but later as free settlements promoted by subsidies given to British people who chose to immigrate to Australia. The discovery of gold in 1851 coincided with an economic depression in Britain, resulting in large-scale immigration from Britain, other parts of Europe, and North America. Workers from China and the Pacific Islands were brought to Australia to labor in mines, farms, and plantations.

The result in Australia was similar to that in South Africa, where distinct groups of people with different genetic and cultural histories were juxtaposed, living with stark social inequality. During the earliest years of European immigration, large numbers of Aboriginal Australians died of infectious disease brought by the settlers, especially smallpox.

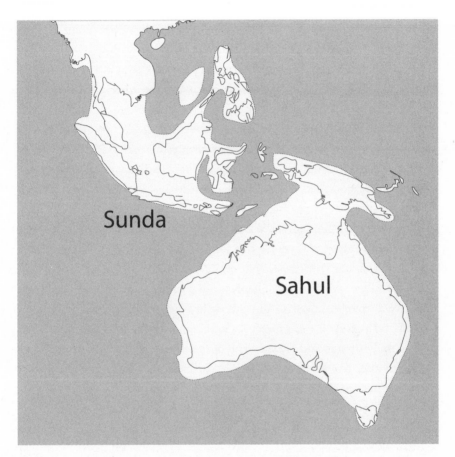

Figure 7.3. The ancient peninsula of Sunda and continent of Sahul during the most recent glacial maximum (ice age), when sea levels were much lower. The approximate locations of ancient shorelines are indicated by dashed lines and the modern shorelines by solid lines.

A conflict named the Black War on the island of Tasmania has been referred to as "the most intense conflict in Australia's history" and "a clash between the most culturally and technologically dissimilar humans to have ever come into contact."[11] A penal colony was established in Tasmania in

1803 and initially remained confined to a small portion of the island. The first conflict was in 1804. By 1820, large numbers of settlers had arrived and encroached on Aboriginal lands as they expanded into the interior. Accounts of what happened vary and are told entirely from the settlers' point of view, and historians continue to dispute what actually happened. It is clear that British colonists considered the Aboriginal people as savages and committed unspeakable violence against them. In response, the Aboriginal attacks on British colonists increased dramatically. Several hundred people lost their lives on both sides. Moreover, infectious disease carried by the colonists took the lives of large numbers of Aboriginal people. In 1828, Governor George Arthur declared martial law, permitting patrols to kill any Aboriginal people who resisted them. After a failed attempt by Arthur in 1830 to round up all Aboriginal people and confine them on a peninsula, George Augustus Robinson, a clergyman and builder, was appointed to be a conciliator and convince the remaining Aboriginal people to relocate to Flinders Island. Robinson succeeded in winning their trust, and, over time, the few hundred people who remained out of what had originally been a population estimated at more than five thousand moved to the island. In spite of Robinson's promises of a good and prosperous life there, living conditions were so poor that most died of malnutrition and disease. The final survivor of the island population (which by then had been moved back to the mainland) died in 1876, and the Tasmanian government announced the extinction of the population. In fact, others had survived elsewhere, and descendants with Aboriginal Tasmanian and European ancestry remain today. Nonetheless, the language, culture, and vast majority of members of an entire people had been annihilated.

On mainland Australia, European immigrants and their descendants viewed Aboriginal Australians and nonwhite, mostly Chinese, immigrants as inferior races. Toward the latter part of the nineteenth century and into the twentieth, fears that the numbers of nonwhite immigrants were becoming too large resulted in what is historically known as the White Australia Policy, a series of laws and political movements aimed at promoting immigration of people recognized as "white" and limiting or excluding immigration of other groups of people. In one of the most famous declarations supporting the White Australia Policy, near the beginning of World War II, Australian prime minister John Curtin stated, "This country shall remain forever the home of the descendants of those people who came here in peace in order to establish in the South Seas an outpost of the British race."[12]

A complex set of discrimination laws and social inequities ensued, including laws governing voting rights, land ownership, antimiscegenation, segregation, and child custody. Among the most infamous is known as the "Stolen Generations." Children with mixed Australian Aboriginal and European ancestry were forcibly removed from their Aboriginal families and placed in institutions or adopted by white families under the assertion that they would lead better lives apart from their families. The practice persisted for a century (1869 through 1970) and affected thousands of children.

Australia's proximity to Japan and the Pacific battles of World War II made it the destination for thousands of Japanese refugees. Some married Australian citizens, and others wished to remain in Australia after the war. Protests over deportation efforts ultimately began to break down the White Australia Policy, resulting in revised laws, such as the Migration Act of 1958, and new policies removing race as a criterion for immigration.[13] Although much has changed in Australia to overcome past racism, social and economic oppression remain as a legacy of historical racism.

We now turn our attention to North and South America, which were the most recent continents to be populated by humans. Toward the end of the last major ice age, when sea levels were still low, a broad land bridge known as Beringia connected Asia and North America (figure 7.4). Though Beringia's climate was cold, much of the coastline was free of ice, providing passage by foot for migrating people. The ancient ancestors of Native Americans entered North America across Beringia about fifteen thousand years ago. According to DNA analysis, they most probably were descendants of populations that resided in central Asia.[14]

Over time, as global temperatures rose and polar ice melted, sea levels climbed, and Beringia was inundated by a frigid and treacherous stretch of seawater. The newly formed Bering Strait separated the Asian and North American continents, effectively isolating human populations in the Americas for thousands of years.

European colonization of the Americas began at the close of the fifteenth century. Settlers from England, France, Sweden, Ireland, Spain, Portugal, the

Netherlands, and, to a lesser extent, other parts of Europe arrived in large numbers over a relatively short period. The Spanish rapidly conquered the vast empires of the Inca, Maya, and Aztec, and the Portuguese established colonies in what is now Brazil. Along the northeast coast of what is now the United States, European (mostly British, Dutch, and German) settlers encountered tribes of Native Americans who lived as both agriculturalists and hunter-gatherers.

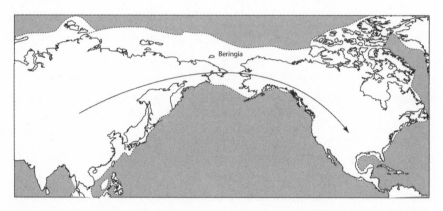

Figure 7.4. The approximate migration route for the ancestors of Native Americans toward the end of the last major ice age, about fifteen thousand years ago.

The stark juxtaposition of European settlers with culturally and genetically different native people that we've seen in the examples of South Africa and Australia was repeated in North America. Native Americans and European immigrants shared common ancestry dating to tens of thousands of years ago, when their ancestors belonged to the same population of people living in the Caucasus region near the Caspian and Black Seas. Their ancestral lines had separated more than thirty thousand years ago, geographically and reproductively, when the ancient ancestors of Europeans migrated westward and the ancient ancestors of Native Americans eastward. They shared common African as well as some postdiaspora variants. Newer variants that differed between them had arisen and accumulated independently since their ancestral separation. The Europeans perceived Native Americans as a separate and inferior race.

As in South Africa and Australia, infectious diseases devastated Native American populations in a series of outbreaks. Though most were inadvertent, at least one was intentional. In an infamous case of germ warfare, Lord Jeffrey Amherst, general commander of the British military during the French and Indian War, mandated that blankets exposed to smallpox victims be given to Native Americans during the conflict known as Pontiac's Rebellion. In Amherst's words from a letter to one of his officers, "You will do well to try to Innoculate the Indians by means of Blanketts, as well as to try every other method that can serve to Extirpate this Execrable Race."[15]

By the eighteenth century, massive transatlantic African slave importation supported the expansion of plantation agriculture in the Americas. Most African slaves were taken from the Atlantic coast of west Africa, extending into the central part of the continent. Because they came mostly from a limited region, the people imported as slaves represented a subset of the human genetic diversity in Africa, mostly from west African Bantu populations. They, too, shared common ancestry with the European and Native Americans, separated by more than sixty thousand years. African slaves constituted some of the largest numbers of immigrants to North America in the eighteenth and nineteenth centuries, numbering nearly four hundred thousand, and more than ten million for all of the Americas.[16] Without freedom or human rights, their descendants lost most of their African cultures, languages, and religions, especially in North America, where they were forced to adopt those of their European American masters.

The westward expansion of European settlement in nineteenth-century North America rapidly accelerated with construction of the US Transcontinental Railroad, accompanied by a major influx of east Asian immigrants to the shores of California to work on the railroad and in mining, agriculture, and other forms of hard labor. Notions of white supremacy were so pervasive at the time that the large number of east Asian immigrants became known as the "yellow menace," "yellow terror," and "yellow peril." National laws in the United States restricting or prohibiting immigration of people with Asian ancestry began with the Page Act of 1875 and continued with the Chinese Exclusion Act of 1882, the Geary Act of 1892, and the "Asiatic Barred Zone" of the Immigration Exclusion Act of 1917.[17]

By the end of the nineteenth century, the United States had sizable populations of people descended from African and European immigrants as well as smaller populations descended from east Asian immigrants. By then, Native Americans had been greatly reduced in number, and many were living on reservations. Although all people at the time were technically free and supposedly equal under the law, the vast majority remained segregated into their particular ethnic groups, with most of the wealth and power concentrated in the European American population.

The westward expansion of people with predominantly European ancestry entered regions that had been part of the Spanish conquest of North America, in a broad region encompassing what is now Texas, California, and the western states. Much of this region had previously been Mexican territory. Several states today retain their Spanish names, such as Colorado (colored), Nevada (snow covered), Montana (mountain), and Arizona (arid zone). Mating of Spanish colonists with Native Americans had been common, and their descendants generations later had settled throughout much of this part of North America. Today, most people with diverse ancestry that includes Native American and Spanish ancestry classify themselves as Hispanic or Latino.

Similar histories could be told for European colonialism nearly everywhere it happened. The perception of discrete races was the overall pattern because distinct populations of people were juxtaposed as a consequence of immigration. When viewed from a worldwide perspective, immigrants did not come from distinct races. However, when viewed from the limited vantage of newly established immigrant colonies, slaves transported from their homelands, and native populations that originally occupied the regions, the presence of distinct races seemed obvious, particularly to colonizers whose preconceived notions of their inherent racial supremacy were embedded in their worldviews, their religious beliefs, and their traditions.

The legacy of this sort of racial categorization remains today. One of the most important points that emerges from that legacy is the distinction between race as a social construct and race as a supposed genetic construct. As we've seen, racial classification makes little sense when viewed in a worldwide genetic context. DNA analysis has confirmed in abundant detail what was partially known from historical and archaeological evidence, which is that complex and massive migration events over the past four thousand years have spread DNA variants in a complex and diverse way throughout most of the people of the world.[18] The overwhelming majority of people on Earth have inherited a mix of variants tracing to major migration events such as these. As one anthropologist put it, "We're all mongrels, we've always been mixing,"[19]

Race as a set of social constructs, however, carries a very different meaning and importance than the notion of races as discrete genetic entities. The social constructs of race differ among nations as political categories because they are based more on the immigration histories of those nations than on worldwide genetic diversity. Recall the statement uttered in 1959 by the judge in the *Loving v. Virginia* antimiscegenation case cited in chapter 1: "Almighty God created the races white, black, yellow, malay and red, and he placed them on separate continents. . . . The fact that he separated the races shows that he did not intend for the races to mix."[20] Such a statement reflects more the perception of race as influenced by US immigration history than the complex genetic histories of people in Europe, Africa, Asia, and the Americas.

A good example of how racial classification is more social than genetic is the racial classification scheme of the US Census. The 2010 census first separated Hispanic classification from race by asking people to classify themselves as Hispanic or non-Hispanic. It then asked all respondents—regardless of their self-classification as Hispanic or non-Hispanic—to further classify themselves into one of the five following racial categories:

White – A person having origins in any of the original peoples of Europe, the Middle East, or north Africa.

Black or African American – A person having origins in any of the Black racial groups of Africa.

American Indian or Alaska Native – A person having origins in any of the original peoples of North and South America (including Central America) and who maintains tribal affiliation or community attachment.

Asian – A person having origins in any of the original peoples of the Far East, Southeast Asia, or the Indian subcontinent including, for example, Cambodia, China, India, Japan, Korea, Malaysia, Pakistan, the Philippine Islands, Thailand, and Vietnam.

Native Hawaiian or Other Pacific Islander – A person having origins in any of the original peoples of Hawaii, Guam, Samoa, or other Pacific Islands.[21]

People who classified themselves into the American Indian or Alaska Native, Asian, and Native Hawaiian or Other Pacific Islander categories were further asked to subclassify themselves. American Indian or Alaska Native respondents were asked to name their tribal affiliation, Asian respondents were asked to name the country of their ancestral origin, and Native Hawaiian or Other Pacific Islander respondents were asked to name the island of their ancestral origin. No such subcategorization was requested for the White and Black categories.

This lack of subcategories for Black and White does not reflect a lack of genetic diversity; people who self-categorize as Black and White ethnic groups in the United States are genetically diverse. Subcategorization in these groups, however, is mostly impossible and has little meaning. Large proportions of people who self-classify into these groups cannot identify a specific country or a specific region for their ancestry, often because some of their immigrant ancestors entered the Americas many generations ago, and their lines of ancestry often trace to various countries. In other categories, however, a relatively large number of people are themselves immigrants or are the recent descendants of immigrants and can thus trace their ancestry to a particular location.

During the twentieth century and into the twenty-first, worldwide immigration from various parts of the world has vastly diversified the populations of many nations. Diversity in the United States, for instance, is far greater now than it was a century ago. And diversity continues to increase. Taboos against antimiscegenation have lost the prominence they once had, although they most certainly have not disappeared. The once-distinct lines of segregation have gradually started to blur.

Race as a social construct is real and meaningful. According to Pilar Ossorio, professor of law and bioethics at the University of Wisconsin and an expert on both social and biological aspects of race,

> Race is deeply rooted in the consciousness of individuals and groups, and it structures our lives and our physical world in myriad ways. It is a strong predictor of where people live, what schools they attend, where and how their spirituality is practiced, what jobs they have, and the amount of income they will earn. Race is real because human beings continually create and recreate it through the process of racialization.[22]

However, race as a social construct is neither universal nor constant. It varies depending on historical, social, and political norms. For instance, apartheid laws in South Africa classified Bantu and Khoisan people into different categories (Black and Coloured, respectively), whereas the US Census defines all people "having origins in any of the Black racial groups of Africa" simply as Black. The social constructs of race can also change over time and across different contexts. As Ossorio explains,

> There is no unitary definition of race, no definition that applies in all places, at all times, and for all purposes. Scholars who include race as a variable in their studies must operationalize the concept of race in a manner that meets the needs of their study, while acknowledging that such "working definitions" merely "fulfill the need for an analytical strategy, they do not reflect a fixed social or biological reality."[23]

Though it may be tempting to promote the utopian ideal of a truly "colorblind" world where race has nothing to do with social, political, or economic status, such an ideal is unrealistic—at least, in today's world and in the foreseeable future. The legacy of past and current racism is powerful and overwhelming. Though enforced racial segregation is no longer legal in the United States, racial distinctions for neighborhoods remain evident in every major US city, and such distinctions are often correlated with economic status. The fact that public schools in the United States are governed and funded largely by geographic location ensures that racial inequality in basic education will persist in spite of efforts targeted at its mitigation. Inequality in employment, public services, healthcare, and many other aspects of society persist and are strongly correlated with racial classification, often as a remnant of past discrimination.

Recognizing how modern perceptions of race arose as artifacts of immigration history, rather than as any sort of definable genetic boundaries or biological basis for race, is essential for understanding what race is and what it is not. Shifting the perception of race away from the notion of a genetic construct and toward the reality of a social construct is critical for what will continue to be a long battle toward eventually defeating racism and purging its legacy.

EPILOGUE

The evidence we've discussed reveals how the human species has evolved since our African origin and how people ultimately spread to occupy the habitable world. The evidence in our DNA shows that genetically we all are strikingly similar to one another. The common chimpanzee, though its natural range covers just a small part of west-central Africa, is genetically more diverse among its populations than are human populations spread across the continents of the world.[1] As humans, we are all closely related—members of the same family, tracing our origins to a common homeland in Africa.

Because its focus is on science, this book has touched only briefly on the history of racism. A full exploration of that history chronicles some of the most unspeakable acts of mass torture and cruelty ever inflicted upon people. The atrocities of centuries-long racial persecution have for too long been sanitized from histories, and many people are unaware of this horrific side of human behavior during the past five centuries. The tide has turned, and a number of well-written, candid, and detailed accounts of the history of racism are now available as books, websites, and documentaries. They deserve a prominent place in historical accounts, complemented by scientific evidence of the type presented here. The history of racism must never be forgotten and should be a principal motivation for the ongoing battle to overcome its legacy.

That history includes a period when proponents of white supremacy used supposed science to promote their cause. A movement now called "scientific racism" was most influential during a period lasting from the mid-nineteenth to the late twentieth century, lingering to some degree even today. Part of its purpose was to preserve the view that nonwhite people were members of distinct races, separate and subordinate to the white race, even to the point where, according to some, nonwhite races did not merit inclusion within the human species. Hypotheses collectively known as *polygenism* posited separate biological origins of different races, an extreme and early version of the multiregional hypothesis of human origins. To some, each race was considered as a separate and distinct species, with only the white race classified as human. Often, proponents mixed Christian theology with their supposedly scientific speculations to claim that Adam and Eve were the original parents of only the white race, other so-called races having allegedly evolved from animals. Under such a scheme, nonwhite people were legally classified as property, just as domestic animals were regarded as property under the law.[2] This belief was one of several used to justify slavery and to deny human rights, and it supported an economic empire of slavery and racial subjugation. Eugenic and antimiscegenation laws, promoted as science, outlasted slavery by a century and often longer, persisting even into the late twentieth century in some places. There are those who still believe that notions of racial purity are biologically and theologically sound, and therefore desirable, in spite of the fact that current genetic evidence has obliterated all justification for such notions.

Some argue that DNA evidence of the type highlighted in this book reinforces rather than invalidates traditional racial classifications. As a recent example, Nicholas Wade, in his book *A Troublesome Inheritance: Genes, Race and Human History*, writes,

> Even when it is not immediately obvious what race a person belongs to from bodily appearance, as may often be the case with people of mixed-race ancestry, race can nonetheless be distinguished at the genomic level. With the help of ancestry informative markers, . . . an individual can be assigned with high confidence to the appropriate continent of origin. If of admixed race, like many African Americans, each block of the genome can be assigned to forebears of African or European ancestry. At least at the level of continental populations, races can be distinguished genetically, and this is sufficient to establish that they exist.[3]

The ability to distinguish African and European DNA segments in African Americans does not imply the existence of discrete races when the distribution of human genetic diversity is considered on a worldwide basis. It reflects, instead, the historical juxtaposition of people whose ancestral backgrounds trace to the discontinuous places of western Africa and northern Europe as a consequence of European colonization and the Atlantic slave trade in North America.

Given the vast amount of human genetic information currently available, traditional racial classifications constitute an oversimplified way to represent the distribution of genetic variation among the people of the world. Mutations have been creating new DNA variants throughout human history, and the notion that a small proportion of them define human races fails to recognize the complex nature of their distribution. A large proportion of variants are very ancient, having arisen more than one hundred thousand years ago in Africa when all people lived there, and are now spread throughout the worldwide human population. These variants have been dispersed over tens of thousands of years as people migrated in myriad ways within Africa, out of Africa, back into Africa, and across the other habitable continents. Other variants display clinal patterns, gradually decreasing in prevalence from a central region. They originated in that region and then were dispersed in people who, over generations, migrated away. Yet other variants tend to be clustered in particular regions, often when barriers such as oceans or mountain ranges inhibited their dispersal during periods of history when people were unable or unlikely to migrate across those barriers. Furthermore, natural selection has resulted in increased prevalence of particular variants in certain parts of the world when those variants conferred an advantage for survival and reproduction. Variants influencing pigmentation or resistance to infectious diseases are prominent examples. In addition, the degree of human genetic diversity varies throughout the world and thus cannot provide a reliable way of classifying humans into particular races on the basis of how much or how little diversity is present. Not surprisingly, genetic diversity is highest by far in sub-Saharan Africa where humans originated. Last, widespread human mobility over the past several millennia has reshuffled the world's genetic diversity in complex ways, negating simplistic preconceptions of so-called pure and mixed races.

The concept of distinct human races as biologically valid groupings was widely accepted prior to the latter part of the twentieth century. It is now outdated, replaced by the more complex and scientifically reliable characterization of genetic ancestry. Some who still cling to the idea of genetically defined races cite evidence that some ancestry informative markers tend to cluster with one another in correlation with geography. Noah Rosenberg of the University of Michigan and his colleagues have conducted some of the most extensive research on such clustering and have made it clear that "our evidence for clustering should not be taken as evidence of our support for any particular concept of 'biological race.'"[4] Notably, some of the world's most prominent human population geneticists have publicly criticized the people who claim genetic research supports the notion of biological races, and the unfounded inferences derived from that notion.[5]

Why should an accurate understanding of our genetic unity and diversity matter? First, there are practical reasons. Variations in health are, in part, a consequence of variations in our DNA. The tendency, however, to associate particular medical conditions with race is often overly simplistic and scientifically flawed. A number of genetic conditions, such as cystic fibrosis and sickle cell anemia, are more prevalent in people whose ancestry traces to particular parts of the world, but conditions such as these are not confined to a particular group, and they are rare in all groups. They should never be labeled as racial diseases.

Dealing with health-related susceptibilities is complex. Genetically inherited differences in skin pigmentation, for example, are inversely correlated with susceptibility to malignant melanoma, the most deadly form of skin cancer. Those with less skin pigmentation are more susceptible than those with greater skin pigmentation. Seemingly racial differences for other susceptibilities, however, may have a much greater social than genetic foundation. Susceptibility to alcoholism in Native Americans, for instance, may have more to do with poverty and poor living conditions than with genetic ancestry.

In this context, race or ethnicity as a social construct can have medical significance. Differences in socioeconomic status, education, quality of healthcare, health practices, substance abuse, infectious disease, and other nongenetic factors may be responsible for at least some of the difference in health conditions among socially defined ethnic groups. In such cases, efforts to overcome poverty, unemployment or underemployment, and educational inequalities are likely to have a commensurate effect on improving overall health.

The same can be said for educational attainment. So-called racial differences in IQ scores are more a consequence of disparities in socioeconomic status and the quality of education than of any genetic differences between ethnic groups. Efforts to improve educational quality and opportunity can increase the economic benefits associated with increased educational achievement. As an educator who often works with first-generation college students, I have seen firsthand the transformation that a well-designed education can generate for those who are economically disadvantaged or lack an adequate educational background.

Beyond the practical reasons, there are perhaps equally important reasons why a scientifically based understanding of human diversity matters. It dispels notions of racial superiority and evokes a sense of wonder and respect for the variety, both genetic and cultural, of the world's human population, from our African origins to the present. Perhaps most important, it tells us who we are and how we originated.

Unfortunately, many people find it difficult to accept what current science tells us about the myth of race. It runs counter to what seem to be obvious racial distinctions, mostly in parts of the world where immigration history has juxtaposed people with discontinuous ancestral backgrounds in the same place. The racial categorizations that many of us have experienced throughout our lives have likewise inculcated a sense of racial division that is not easy to abandon. Regardless of what the scientific evidence shows, the perception of race and the associated racial discrimination are unlikely to disappear anytime soon. Furthermore, a scientific understanding of human evolutionary history challenges commonly held religious beliefs that are based on literal interpretations of biblical history. Everything we have discussed in this book, and everything related to human genetic diversity, is a consequence of our evolutionary history, supported by abundant evidence. Just as geneticist Theodosius Dobzhansky famously affirmed that "nothing in biology makes sense except in the light of evolution," nothing we know about human genetic diversity makes sense except in the light of human evolution.[6] Yet a significant minority of people (42 percent in the United States, according to the most recent Gallup poll) fully reject human evolution, opting instead for the belief that humans were specially created with no prior evolutionary ancestry less than ten thousand years ago.[7] Such beliefs are often infused with a nonscientific perception of different races and how they supposedly originated. And, in spite of overwhelming scientific evidence and changing social norms, a relatively large proportion of people still cling to past traditions of white supremacy and racism.

To understand who we are as a species, and why we vary as we do, we must examine our genetic diversity in the context of a common African origin, followed by intra- and intercontinental diasporas that transpired over a period of tens of thousands of years, culminating in an era of major migrations that reshuffled the worldwide human genetic constitution over the past several thousand years and is still underway. Last, we must recognize that today's human population is far larger, more diverse, and more complex than it ever has been. We are all related, more than seven billion of us, distant cousins to one another, and, ultimately, everyone is African.

NOTES

PREFACE

1. R. J. Sternberg, "Intelligence," *Dialogues in Clinical Neuroscience* 14, no. 1 (2012): 19–27.

2. British Broadcasting Corporation (BBC), "Episode 3: A Savage Legacy," *Racism: A History*, 58:47, http://topdocumentaryfilms.com/racism-history (accessed June 25, 2014), 58:02.

3. Ibid., 55:40.

4. A. James, "Making Sense of Race and Racial Classification," in *White Logic, White Methods: Racism and Methodology*, ed. T. Zuberi et al. (Lanham, MD: Rowman and Littlefield, 2008), p. 32.

5. BBC, *Racism: A History*, http://topdocumentaryfilms.com/racism-history (accessed June 25, 2014); Public Broadcasting System, *Race: The Power of an Illusion*, http://www.pbs.org/race (accessed June 25, 2014).

CHAPTER 1: WHAT IS RACE?

1. As quoted by E. Warren, "*Loving v. Virginia*: Opinion of the Court," No. 395, 206 Va. 924, 147 S.E.2d 78, reversed, http://www.law.cornell.edu/supct/html/historics/USSC_CR_0388_0001_ZO.html (accessed July 15, 2014).

2. Ibid.

3. Ibid.

4. T. Head, "Interracial Marriage Laws: A Short Timeline History," http://civilliberty.about.com/od/raceequalopportunity/tp/Interracial-Marriage-Laws-History-Timeline.htm (accessed May 14, 2013).

5. South Africa Parliament, *Report of the Joint Committee on the Prohibition of Mixed Marriages Act and Section 16 of the Immorality Act* (Cape Town, South Africa: Government Printer, 1985).

6. C. R. Darwin, *On the Origin of Species by Means of Natural Selection, or the Preservation of Favoured Races in the Struggle for Life*, 4th ed. (London: John Murray, 1866), p. 16.

7. C. R. Darwin, *On the Origin of Species by Means of Natural Selection, or the Preservation of Favoured Races in the Struggle for Life*, 5th ed. (London: John Murray, 1869), p. 243.

8. American Kennel Club, "Breed Matters," https://www.akc.org/breeds (accessed May 3, 2014).

9. C. R. Darwin, *On the Origin of Species by Means of Natural Selection, or the Preservation of Favoured Races in the Struggle for Life*, 1st ed. (London: John Murray, 1859), p. 298.

10. R. C. Punnett, *Mendelism* (New York: Macmillan, 1905), p. 184.

11. United States Holocaust Memorial Museum, "Holocaust Encyclopedia," http://www.ushmm.org/wlc/en/article.php?ModuleId=10005143 (accessed July 16, 2012).

12. Jewish Virtual Library, "The Nazi Party: The 'Lebensborn' Program," http://www.jewishvirtuallibrary.org/jsource/Holocaust/Lebensborn.html (accessed July 16, 2012).

13. A. M. Stern, *Eugenic Nation: Faults and Frontiers of Better Breeding in Modern America* (Oakland, CA: University of California Press, 2005), p. 244.

14. G. Hellenthal et al., "A Genetic Atlas of Human Admixture History," *Science* 343, no. 6172 (2014): 747–51.

15. R. C. Lewontin, "The Apportionment of Human Diversity," *Evolutionary Biology* 6 (1972): 385.

16. Ibid., p. 382.

17. J. P. Jarvis et al., "Patterns of Ancestry, Signatures of Natural Selection, and Genetic Association with Stature in Western African Pygmies," *PLoS Genetics* 8, no. 4 (2012): e1002641.

18. Ibid.

19. S. E. Lederer, *Flesh and Blood: Organ Transplantation and Blood Transfusion in 20th Century America* (Oxford: Oxford University Press, 2008).

20. A. W. F. Edwards, "Human Genetic Diversity: Lewontin's Fallacy," *BioEssays* 25, no. 8 (2003): 800.

21. Ibid., p. 801.

22. L. B. Jorde and S. P. Wooding, "Genetic Variation, Classification, and 'Race,'" *Nature Genetics* 36 (2004): S28.

23. Ibid., p. S30.

CHAPTER 2: AFRICAN ORIGINS

1. The evidence of this ancient population is human skeletal remains, including nearly intact skulls bearing the features of modern humans, in the Qafzeh and Skhul caves. See C. B. Stringer et al., "ESR Dates for the Hominid Burial Site of Es Skhul in Israel," *Nature* 338 (1989): 756–58.

2. D. J. Fairbanks, *Evolving: The Human Effect and Why It Matters* (Amherst, NY: Prometheus Books, 2012).

3. Ibid.

4. R. E. Green et al., "A Draft Sequence of the Neanderthal Genome," *Science* 328, no. 7929 (2010): 710–22.

5. J. Zhang et al., "Genomewide Distribution of High-Frequency, Completely Mismatching SNP Haplotype Pairs Observed to Be Common across Human Populations," *American Journal of Human Genetics* 73, no. 5 (2003): 1073–81.

6. L. B. Jorde and S. P. Wooding, "Genetic Variation, Classification, and Race," *Nature Genetics* 36 (2004): S28–S33.

7. There is one documented instance of paternal inheritance of mitochondrial DNA in humans, and it is due to a genetic disorder. The man who inherited this DNA had the paternal mitochondrial DNA only in his muscles. The mitochondrial DNA in the rest of his body was maternal, so this single documented instance of paternal transmission of mitochondrial DNA has no effect on pure maternal inheritance throughout generations. The research was published by M. Schwartz and D. Vissing, "Paternal Inheritance of Mitochondrial DNA," *New England Journal of Medicine* 347, no. 8 (2002): 576–80.

8. M. Ingman et al., "Mitochondrial Genome Variation and the Origin of Modern Humans," *Nature* 408, no. 6828 (2000): 708–13; N. van Oven and M. Kayser, "Updated Comprehensive Phylogenetic Tree of Global Human Mitochondrial DNA Variation," *Human Mutation* 30, no. 2 (2009): E386–94; P. Soares et al., "Correcting for Purifying Selection: An Improved Human Mitochondrial Molecular Clock," *American Journal of Human Genetics* 84, no. 6 (2009): 740–59.

9. U. A. Perego et al., "Distinctive Paleo-Indian Migration Routes from Beringia Marked by Two Rare mtDNA Haplogroups," *Current Biology* 13, no. 1 (2009): 1–8.

10. S. A. Elias, "Late Pleistocene Climates of Beringia, Based on Analysis of Fossil Beetles," *Quaternary Research* 53, no. 2 (2000): 229–35.

11. B. Malyarchuk et al., "The Peopling of Europe from the Mitochondrial Haplogroup U5 Perspective," *PLoS ONE* 5, no. 4 (2010): e10285.

12. D. M. Behar et al., "A 'Copernican' Reassessment of the Human Mitochondrial DNA Tree from Its Root," *American Journal of Human Genetics* 90, no. 4 (2012): 675–84.

13. Soares et al., "Correcting for Purifying Selection."

14. In reality, a very small part on one end of the Y chromosome recombines with the X chromosome. However, all genetic analysis for ancestry is done with the nonrecombining portion, which represents the vast majority of the Y chromosome.

15. F. Cruciani et al., "A Revised Root for the Human Y Chromosomal Phylogenetic Tree: The Origin of Patrilineal Diversity in Africa," *American Journal of Human Genetics* 88, no. 6 (2011): 814–18.

16. W. Fu et al., "Analysis of 6,515 Exomes Reveals the Recent Origin of Most Human Protein-Coding Variants," *Nature* 493, no. 7431 (2013): 216–20.

17. Jorde and Wooding, "Genetic Variation, Classification, and Race," p. S29.

CHAPTER 3: ANCESTRY VERSUS RACE

1. For an excellent summary of the Jefferson-Hemings history and research, see the PBS *Frontline* episode "Mapping Jefferson's Y Chromosome" at http://www. pbs.org/wgbh/pages/frontline/shows/jefferson/etc/genemap.html (accessed November 11, 2012). For more detailed information, and the results of original research, see F. L. Mendez et al., "Increased Resolution of Y Chromosome Haplogroup T Defines Relationships among Populations of the Near East, Europe, and Africa," *Human Biology* 83, no. 1 (2011): 39–53; Thomas Jefferson Memorial Foundation, *Report of the Research Committee on Thomas Jefferson and Sally Hemings*, http://www.monticello .org/sites/default/files/inline-pdfs/jefferson-hemings_report.pdf (accessed November 11, 2012); National Public Radio, "Thomas Jefferson Descendants Work to Heal Family's Past," http://www.npr.org/templates/story/story.php?storyId=131243217 (accessed November 11, 2012); A. G. Reed, *The Hemingses of Monticello: An American Family* (New York: W. W. Norton, 2009).

2. D. M. Goldenberg, *The Curse of Ham: Race and Slavery in Early Judaism, Christianity, and Islam* (Princeton, NJ: Princeton University Press, 2003).

3. Ibid.

4. S. J. Gould, *The Mismeasure of Man*, rev. ed. (New York: W. W. Norton, 1996), p. 404.

5. United States Census Bureau, "Race," http://www.census.gov/topics/ population/race.html (accessed June 25, 2014).

6. M. F. Hammer et al., "Population Structure of Y Chromosome SNP Haplogroups in the United States and Forensic Implications for Constructing Y Chromosome STR Databases," *Forensic Science International* 164, no. 1 (2006): 45–55.

7. D. J. Fairbanks et al., "*NANOGP8*: Evolution of a Human-Specific Retro-Oncogene," *G3: Genes Genomes Genetics* 2, no. 11 (2012): 1447–57; D. J. Fairbanks and P. J. Maughan, "Evolution of the *NANOG* Pseudogene Family in the Human and Chimpanzee Genomes," *BMC Evolutionary Biology* 6 (2006): 12.

8. I. Chambers et al., "Functional Expression Cloning of *Nanog*, a Pluripotency Sustaining Factor in Embryonic Stem Cells," *Cell* 113, no. 5 (2003): 643–55.

9. L. Ségurel et al., "The ABO Blood Group Is a Trans-Species Polymorphism in Primates," *Proceedings of the National Academy of Sciences, USA* 109, no. 45 (2012): 18493–98.

10. J. A. Rowe et al., "Blood Group O Protects against Severe *Plasmodium falciparum* Malaria through the Mechanism of Reduced Rosetting," *Proceedings of the National Academy of Sciences, USA* 104, no. 44 (2007): 17471–76; A. E. Fry et al., "Common Variation in the ABO Glycosyltransferase Is Associated with Susceptibility to Severe *Plasmodium falciparum* Malaria," *Human Molecular Genetics* 17, no. 4 (2008): 567–76.

11. R. I. Glass et al., "Predisposition for Cholera of Individuals with O Blood Group: Possible Evolutionary Significance," *American Journal of Epidemiology* 121, no. 6 (1985): 791–96; J. D. Clemens et al., "ABO Blood Groups and Cholera: New Observations on Specificity of Risk and Modification of Vaccine Efficacy," *Journal of Infectious Disease* 159, no. 4 (1989): 770–73; A. S. G. Faruque et al., "The Relationship between ABO Blood Groups and Susceptibility to Diarrhea due to *Vibrio cholerae* 0139," *Clinical Infectious Disease* 18, no. 5 (1994): 827–28.

12. A. Keinan and A. G. Clark, "Recent Explosive Human Population Growth Has Resulted in an Excess of Rare Genetic Variants," *Science* 336, no. 6082 (2012): 740–43.

13. Fairbanks et al., "*NANOGP8*," pp. 1447–57.

14. J. Xing et al., "Fine-Scaled Human Genetic Structure Revealed by SNP Microarrays," *Genome Research* 19 (2009): 819.

15. E. Giardina et al., "Haplotypes in *SLC24A5* Gene as Ancestry Informative Markers in Different Populations," *Current Genomics* 9, no. 2 (2008): 110–14.

16. Race, Ethnicity, and Genetics Working Group, "The Use of Racial, Ethnic, and Ancestral Categories in Human Genetics Research," *American Journal of Human Genetics* 77, no. 4 (2005): 524.

CHAPTER 4: "THE COLOR OF THEIR SKIN"

1. M. L. King Jr., "I Have a Dream," *Historic Documents*, http://www.ushistory.org/documents/i-have-a-dream.htm (accessed June 29, 2014).

2. C. R. Darwin, *On the Origin of Species by Means of Natural Selection or the Preservation of Favoured Races in the Struggle for Life*, 1st ed. (London: John Murray, 1859), p. 406.

3. N. G. Jablonski and G. Chaplin, "Human Skin Pigmentation as an Adaptation to UV Radiation," *Proceedings of the National Academy of Sciences, USA* 107, suppl. 2 (2010): 8962–68.

4. G. S. Omenn, "Evolution and Public Health," *Proceedings of the National Academy of Sciences, USA* 107, suppl. 1 (2012): 1702–709; S. Sharma et al., "Vitamin D Deficiency and Disease Risk among Aboriginal Arctic Populations," *Nutritional Review* 69, no. 8 (2011): 468–78.

5. C. R. Wagner, F. R. Greer, and the Section on Breastfeeding and Committee on Nutrition, "Prevention of Rickets and Vitamin D Deficiency in Infants, Children, and Adolescents," *Pediatrics* 122, no. 5 (2008): 1142–52.

6. T. Haitina et al., "High Diversity in Functional Properties of Melanocortin 1 Receptor (*MC1R*) in Divergent Primate Species Is More Strongly Associated with Phylogeny than Coat Color," *Molecular Biology and Evolution* 24, no. 9 (2007): 2001–8.

7. P. Sulem et al., "Genetic Determinants of Hair, Eye and Skin Pigmentation in Europeans," *Nature Genetics* 39 (2007): 1443–52; E. Po piech et al., "The Common Occurrence of Epistasis in the Determination of Human Pigmentation and Its Impact on DNA-Based Pigmentation Phenotype Prediction," *Forensic Science International: Genetics* 11 (2014): 64–72.

8. S. Beleza et al., "The Timing of Pigmentation Lightening in Europeans," *Molecular Biology and Evolution* 30, no. 1 (2013): 24–35.

9. E. Healy et al., "Functional Variation of *MC1R* Alleles from Red-Haired Individuals," *Human Molecular Genetics* 10, no. 21 (2001): 2397–402.

10. Jablonski and Chaplin, "Human Skin Pigmentation"; K. Makova and H. L. Norton, "Worldwide Polymorphism at the *MC1R* Locus and Normal Pigmentation Variation in Humans," *Peptides* 26, no. 10 (2005): 1901–908.

11. Jablonski and Chaplin, "Human Skin Pigmentation."

12. S. J. Gould, *The Mismeasure of Man*, rev. ed. (New York: W. W. Norton, 1996), p. 401.

13. E. Giardina et al., "Haplotypes in *SLC24A5* Gene as Ancestry Informative Markers in Different Populations," *Current Genomics* 9, no. 2 (2008): 110–14.

14. R. Smith et al., "Melanocortin 1 Receptor Variants in an Irish Population," *Journal of Investigative Dermatology* 111, no. 1 (1998): 119–22.

15. C. Lalueza-Fox et al., "A Melanocortin 1 Receptor Allele Suggests Varying Pigmentation among Neanderthals," *Science* 318, no. 5855 (2007): 1453–55.

16. C. C. Cequeira et al., "Predicting *Homo* Pigmentation Phenotype through Genomic Data: From Neanderthal to James Watson," *American Journal of Human Biology* 24, no. 5 (2012): 705–709.

17. Jablonsky and Chaplin, "Human Skin Pigmentation," p. 8966.

18. H. L. Norton et al., "Genetic Evidence for the Convergent Evolution of Light Skin in Europeans and East Asians," *Molecular Biology and Evolution* 24, no. 3 (2007): 710–22.

CHAPTER 5: HUMAN DIVERSITY AND HEALTH

1. A. C. Allison, "Protection Afforded by Sickle-Cell Trait against Subtertian Malarial Infection," *British Medical Journal* 1, no. 4857 (1954): 290–94.

2. S. M. Rich et al., "The Origin of Malignant Malaria," *Proceedings of the National Academy of Sciences, USA* 106, no. 35 (2009): 14902–907.

3. M. Currat et al., "Molecular Analysis of the Beta-Globin Gene Cluster in the Niokholo Mandenka Population Reveals a Recent Origin of the Beta-S Senegal Mutation," *American Journal of Human Genetics* 70, no. 1 (2002): 207–23.

4. A. E. Kulozik et al., "Geographical Survey of Beta-S-Globin Gene Haplotypes: Evidence for an Independent Asian Origin of the Sickle-Cell Mutation," *American Journal of Human Genetics* 39, no. 2 (1986): 239–44; F. Y. Zeng et al., "Sequence of the −530 Region of the Beta-Globin Gene of Sickle Cell Anemia Patients with the Arabian Haplotype," *Human Mutation* 3, no. 2 (1994): 163–65.

5. American Society of Hematology, "Statement on Screening for Sickle Cell Trait and Athletic Participation," http://www.hematology.org/Advocacy/Statements/2650.aspx (accessed June 25, 2014).

6. Ibid.

7. Sickle Cell Disease Association of America, "Sickle Cell Trait and Athletics," http://www.sicklecelldisease.org/index.cfm?page=sickle-cell-trait-athletics (accessed December 30, 2012).

8. J. C. Goldsmith et al., "Framing the Research Agenda for Sickle Cell Trait: Building on the Current Understanding of Clinical Events and their Potential Implications," *American Journal of Hematology* 87, no. 3 (2012): 340–46.

9. N. L. Kaplan et al., "Age of the Δ*F508* Cystic Fibrosis Mutation," *Nature Genetics* 8 (1994): 216–18; N. Morral et al., "The Origin of the Major Cystic Fibrosis Mutation (Δ*F508*) in European Populations," *Nature Genetics* 7, no. 2 (1994): 169–75; E. Mateu et al., "Can a Place of Origin of the Main Cystic Fibrosis Mutations Be Identified?" *American Journal of Human Genetics* 70, no. 1 (2002): 257–64.

10. The data cited here are from the US Centers for Disease Control and Prevention (CDC), "U. S. Cancer Statistics: An Interactive Atlas," http://apps.nccd .cdc.gov/DCPC_INCA/DCPC_INCA.aspx (accessed January 12, 2013).

11. CDC, "Adult Cigarette Smoking in the United States: Current Estimates," http://www.cdc.gov/tobacco/data_statistics/fact_sheets/adult_data/cig _smoking/#state (accessed January 12, 2013).

12. C. Lu, "What Causes "Asian Glow?" *Yale Scientific*, April 3, 2011, http:// www.yalescientific.org/2011/04/what-causes-"asian-glow" (accessed January 5, 2013).

13. Y. Matsuo, R. Yokoyama, and S. Yokoyama, "The Genes for Human Alcohol Dehydrogenases Beta-1 and Beta-2 Differ by Only One Nucleotide," *European Journal of Biochemistry* 183, no. 2 (1989): 317–20.

14. H. Li et al., "Ethnic Related Selection for an ADH Class I Variant within East Asia," *PLoS One* 3, no. 4 (2008): 2.

15. Ibid.

16. J. Liu et al., "Haplotype-Based Study of the Association of Alcohol Metabolizing Genes with Alcohol Dependence in Four Independent Populations," *Alcoholism: Clinical and Experimental Research* 35, no. 2 (2011): 304–16.

17. C. Hedges and J. Sacco, *Days of Destruction, Days of Revolt* (New York: Nation Books, 2012), pp. 2–3.

18. C. L. Ehlers, "Variations in ADH and ALDH in Southwest California Indians," *Alcohol Research & Health* 30, no. 1 (2007): 14–17.

19. C. L. Ehlers et al., "Linkage Analyses of Stimulant Dependence, Craving and Heavy Use in American Indians," *American Journal of Medical Genetics Part B Neuropsychiatric Genetics* 156B, no. 7 (2011): 772–80.

20. T. Nakajima et al., "Natural Selection and Population History in the Human Angiotensinogen Gene (*AGT*): 736 Complete *AGT* Sequences in Chromosomes from around the World," *American Journal of Human Genetics* 74, no. 5 (2004): 898–916.

21. T. L. Savittand and M. F. Goldberg, "Herrick's 1910 Case Report of Sickle Cell Anemia. The Rest of the Story," *Journal of the American Medical Association* 261, no. 2 (1989): 266–71.

22. L. Pauling et al., "Sickle Cell Anemia: A Molecular Disease," *Science* 110 (1949): 543–48.

23. G. S. Graham and S. H. McCarty, "Sickle Cell (Meniscocytic) Anemia," *Southern Medical Journal* 23 (1930): 600, quoted in M. Tapper, *In the Blood: Sickle Cell Anemia and the Politics of Race* (Philadelphia: University of Pennsylvania Press, 1999), p. 38.

24. M. F. Hammer et al., "Population Structure of Y Chromosome SNP Haplogroups in the United States and Forensic Implications for Constructing Y Chromosome STR Databases," *Forensic Science International* 164, no. 1 (2006): 45–55.

25. National Humanities Center Resource Toolbox, "On Slaveholders' Sexual Abuse of Slaves: Selections from 19th & 20th Century Slave Narratives," *The Making of African American Identity, Vol. I, 1500–1865*, http://nationalhumanitiescenter.org/pds/maai/enslavement/text6/masterslavesexualabuse.pdf (accessed January 15, 2013).

26. A. B. Raper, "Sickle Cell Disease in Africa and America: A Comparison," *Journal of Tropical Medicine and Hygiene* 53 (1950): 53, quoted in Tapper, *In the Blood*, p. 41.

27. R. B. Scott, "Health Care Priorities and Sickle Cell Anemia," *Journal of the American Medical Association* 214, no. 4 (1970): 731, quoted in Tapper, *In the Blood*, p. 102.

28. Tapper, *In the Blood*, p. 104.

29. Pauling et al., "Sickle Cell Anemia."

30. R. B. Scott, "Health Care Priorities and Sickle Cell Anemia," p. 734, quoted in Tapper, *In the Blood*, pp. 105–106.

31. R. B. Scott, "Reflections on the Current Status of the National Sickle Cell Disease Program in the United States," *Journal of the National Medical Association* 71, no. 7 (1979): 679–81.

32. S. A. Tishkoff et al., "Convergent Adaptation of Human Lactase Persistence in Africa and Europe," *Nature Genetics* 39 (2007): 31–40.

33. S. H. Witt, "Pressure Points in Growing up Indian," *Perspectives* 12, no. 1 (1980): 24–31.

34. Ibid.

35. M. S. Watson et al., eds., "Newborn Screening: Toward a Uniform Screening Panel and System," *Genetics in Medicine* 8, suppl. 1 (2006): 1s–252s.

36. B. M. Rusert and C. D. M. Royal, "Grassroots Marketing in a Global Era: More Lessons from BiDil," *Journal of Law and Medical Ethics* 39, no. 1 (2011): 79–90.

37. Ibid.

38. P. C. Ng et al., "Individual Genomes Instead of Race for Personalized Medicine," *Clinical Pharmacology and Therapeutics* 84 (2008): 306.

39. H. Brody and L. M. Hunt, "BiDil: Assessing a Race-Based Pharmaceutical," *Annals of Family Medicine* 4, no. 6 (2006): 558.

40. Ibid., p. 559.

41. L. K. Williams et al., "Differing Effects of Metformin on Glycemic Control by Race-Ethnicity," *Journal of Clinical Endocrinology and Metabolism* (early release, in press, 2014), http://press.endocrine.org/doi/abs/10.1210/jc.2014-1539 (accessed June 25, 2014).

42. A. Wojcicki, "23andMe Provides an Update Regarding FDA's Review," *23andMe Blog*, December 5, 2013, http://blog.23andme.com/news/23andme -provides-an-update-regarding-fdas-review (accessed January 5, 2013).

43. A. Jolie, "My Medical Choice," *New York Times*, May 14, 2013, http://www .nytimes.com/2013/05/14/opinion/my-medical-choice.html?_r=0 (accessed January 5, 2014).

44. P. R. Billings et al., "Discrimination as a Consequence of Genetic Testing," *American Journal of Human Genetics* 50, no. 3 (1992): 476–82.

45. Coalition for Genetic Fairness, "The History of GINA," http://www.genetic alliance.org/ginaresource.history (accessed January 27, 2013).

CHAPTER 6: HUMAN DIVERSITY AND INTELLIGENCE

1. S. J. Gould, *The Mismeasure of Man* (New York: W. W. Norton, 1981).

2. S. J. Gould, *The Mismeasure of Man*, rev. ed. (New York: W. W. Norton, 1996).

3. R. J. Herrnstein and C. Murray, *The Bell Curve: Intelligence and Class Structure in American Life* (New York: Free Press, 1996).

4. Gould, *Mismeasure of Man*, rev. ed., p. 368.

5. C. F. Chabris, "IQ Since 'The Bell Curve,'" *Commentary* 106 (1998): 33–40, http://www.wjh.harvard.edu/~cfc/Chabris1998a.html (accessed February 5, 2013).

6. A. R. Jensen, "How Much Can We Boost IQ and Scholastic Achievement?" *Harvard Educational Review* 39, no. 2 (1969): 1–123.

7. J. P. Rushton and A. R. Jensen, "Thirty Years of Research on Race Differences in Cognitive Ability," *Psychology, Public Policy, and Law* 11, no. 2 (2005): 235.

8. Gould, *Mismeasure of Man*, rev. ed., p. 369.

9. Ibid.

10. As quoted in a promotional statement on the first page of Herrnstein and Murray, *Bell Curve*.

11. L. Hodges, "The Bell Curve Is Sending Shock Waves through America," http://www.timeshighereducation.co.uk/story.asp?storyCode=154396&sectioncode =26 (accessed February 1, 2013).

12. Gould, *Mismeasure of Man*, rev. ed., pp. 376–77.

13. L. S. Gottfredson, "Mainstream Science on Intelligence: An Editorial with 52 Signatories, History, and Bibliography," *Intelligence* 24, no. 1 (1997): 13–23.

14. U. Neisser et al., "Intelligence: Knowns and Unknowns," *American Psychologist* 51 (1996): p. 77.

15. Ibid.

16. R. E. Nisbett et al., "Intelligence: New Findings and Theoretical Developments," *American Psychologist* 67, no. 2 (2012): 130–59.

17. Herrnstein and Murray, *Bell Curve*, p. 318.

18. Ibid., p. 276.

19. Neisser et al., "Intelligence: Knowns and Unknowns"; Nisbett et al., "Intelligence: New Findings."

20. Herrnstein and Murray, *Bell Curve*, p. 276.

21. Ibid., pp. 298–99.

22. Ibid., p. 311.

23. Rushton and Jensen, "Race Differences in Cognitive Ability."

24. Ibid., pp. 265–66.

25. L. S. Gottfredson, "What If the Hereditarian Hypothesis Is True?" *Psychology, Public Policy, and Law* 11, no. 2 (2005): 316.

26. R. E. Nisbett, "Heredity, Environment, and Race Differences in IQ: A Commentary on Rushton and Jensen," *Psychology, Public Policy, and Law* 11, no. 2 (2005): 302.

27. R. J. Sternberg, "There Are No Public-Policy Implications: A Reply to Rushton and Jensen," *Psychology, Public Policy, and Law* 11, no. 2 (2005): 295.

28. R. J. Sternberg, "Intelligence," *Dialogues in Clinical Neuroscience* 14, no. 1 (2012): 24.

29. C. S. Spearman, "'General Intelligence,' Objectively Determined and Measured," *American Journal of Psychology* 15, no. 2 (1904): 201–92.

30. Sternberg, "Intelligence," p. 21.

31. R. J. Sternberg, E. L. Grigorenko, and K. K. Kidd, "Intelligence, Race, and Genetics," *American Psychologist* 60, no. 2 (2005): 47.

32. Nisbett et al., "Intelligence: New Findings," p. 131.

33. Ibid.

34. Sternberg, "Intelligence."

35. R. C. Lewontin, "Race and Intelligence," *Bulletin of the Atomic Scientists* 26 (1970): 2–8.

36. Nisbett et al., "Intelligence: New Findings," p. 132.

37. Herrnstein and Murray, *Bell Curve*, p. 105.

38. Ibid., p. 132.

39. Ibid., p. 107.

40. Nisbett et al., "Intelligence: New Findings."

41. R. Plomin, "Child Development and Molecular Genetics: 14 Years Later," *Child Development* 84, no. 1 (2013): 104–20.

42. J. R. Flynn, *What Is Intelligence? Beyond the Flynn Effect* (Cambridge: Cambridge University Press, 2007), p. 2.

43. Nisbett et al., "Intelligence: New Findings."

44. Ibid.

45. T. C. Daley et al., "IQ on the Rise: The Flynn Effect in Rural Kenyan Children," *Psychological Science* 14, no. 3 (2003): 215–19; G. Meisenberg et al., "The Flynn Effect in the Caribbean: Generational Change in Test Performance in Dominica," *Mankind Quarterly* 46 (2005): 29–70.

46. Nisbett et al., "Intelligence: New Findings," p. 140.

47. Ibid., p. 141.

48. Ibid.

49. R. Plomin and M. Rutter, "Child Development, Molecular Genetics, and What to Do with Genes Once They Are Found," *Child Development* 69, no. 4 (1998): 1223–42.

50. R. Plomin, "Child Development and Molecular Genetics: 14 Years Later," *Child Development* 84, no. 1 (2013): 104.

51. M. Trzaskowski et al., "DNA Evidence for Strong Genetic Stability and Increasing Heritability of Intelligence from Age 7 to 12," *Molecular Psychiatry* 19, no. 3 (2014): 380–84.

52. C. F. Chabris et al., "Most Reported Genetic Associations with General Intelligence Are Probably False Positives," *Psychological Science* 23, no. 11 (2011): 1314–23.

53. B. Benyamin et al., "Childhood Intelligence Is Heritable, Highly Polygenic and Associated with *FNBP1L*," *Molecular Psychiatry* 19, no. 2 (2014): 253–58; G. Davies et al., "Genome-Wide Association Studies Establish That Human Intelligence Is Highly Heritable and Polygenic," *Molecular Psychiatry* 6, no. 10 (2011): 996–1005.

54. Gould, *Mismeasure of Man*, rev. ed., p. 187.

55. Benyamin et al., "Childhood Intelligence Is Heritable"; Davies et al., "Genome-Wide Association Studies."

56. For instance, using an analogy of growing corn in Iowa as opposed to the Mojave Desert to illustrate environmental differences, Herrnstein and Murray write in *The Bell Curve*, "The environment for American Blacks has been closer to the Mojave and the environment for American whites has been closer to Iowa" (p. 298).

CHAPTER 7: THE PERCEPTION OF RACE

1. G. Hellenthal et al., "A Genetic Atlas of Human Admixture History," *Science* 343, no. 6172 (2014): 747–51.

2. B. M. Henn et al., "Genomic Ancestry of North Africans Supports Back-to-Africa Migrations," *PLoS Genetics* 8, no. 1 (2012): e1002397.

3. "Companion website for 'A Genetic Atlas of Human Admixture History,'" http://admixturemap.paintmychromosomes.com (accessed February 23, 2014).

4. All dates for mitochondrial divergences listed in this chapter are from Soares et al., "Correcting for Purifying Selection: An Improved Human Mitochondrial Molecular Clock," *American Journal of Human Genetics* 84, no. 6 (2009): 740–59.

5. A. Zoutendyk, A. C. Kopec, and A. E. Mourant, "The Blood Groups of the Hottentots," *American Journal of Physical Anthropology* 13, no. 4 (1955): 691–97.

6. South African History Online, "The Battle of Blood River," http://www.sahistory.org.za/dated-event/battle-blood-river (accessed April 28, 2014).

7. G. Chin and E. Culotta, "The Science of Inequality," *Science* 344 (2014): 819–21.

8. M. Rassmussen et al., "An Aboriginal Australian Genome Reveals Separate Human Dispersals into Asia," *Science* 334 (2011): pp. 94–98.

9. G. R. Summerhayes et al., "Human Adaptation and Plant Use in Highland New Guinea 49,000 to 44,000 Years Ago," *Science* 330, no. 6186 (2010): 78–81.

10. C. N. Johnson and B. W. Brook, "Reconstructing the Dynamics of Ancient Human Populations from Radiocarbon Dates: 10,000 Years of Population Growth in Australia," *Proceedings of the Royal Society, Series B* 278 (2011): 3748–54.

11. Clements, N, "The Tasmanian Black War: A Tragic Case Lest We Remember?" *The Conversation*, http://theconversation.com/tasmanias-black-war-a-tragic-case-of-lest-we-remember-25663 (accessed June 22, 2014).

12. Australian Government, Department of Immigration and Border Protection,

"Fact Sheet 8. Abolition of the 'White Australia' Policy," http://www.immi.gov.au/media/fact-sheets/08abolition.htm (accessed April 19, 2014).

13. Ibid.

14. M. Raghavan et al., "Upper Palaeolithic Siberian Genome Reveals Dual Ancestry of Native Americans," *Nature* 505 (2014): 87–91; M. C. Dulik et al., "Mitochondrial DNA and Y Chromosome Variation Provides Evidence for a Recent Common Ancestry between Native Americans and Indigenous Altaians," *American Journal of Human Genetics* 90, no. 2 (2012): 229–46; M. V. Derenko et al., "The Presence of Mitochondrial Haplogroup X in Altaians from South Siberia," *American Journal of Human Genetics* 69, no. 1 (2001): 237–41.

15. For a photograph of the document, see http://people.umass.edu/derrico/amherst/34_41_114_fn.jpeg (accessed April 27, 2014). See also P. D'Errico, "Jeffrey Amherst and Smallpox Blankets," http://people.umass.edu/derrico/amherst/lord_jeff .html (accessed April 27, 2014).

16. "Voyages," *Trans Atlantic Slave Trade Database*, http://www.slavevoyages.org/tast/index.faces (accessed December 24, 2013).

17. For texts of the original documents, see http://library.uwb.edu/guides/usimmigration/18%20stat%20477.pdf, http://www.ourdocuments.gov/doc.php ?flash=true&doc=47, http://www.sanfranciscochinatown.com/history/1892gearyact .html, http://library.uwb.edu/guides/usimmigration/39%20stat%20874.pdf (accessed May 3, 2014).

18. Except where otherwise noted, the examples described in this paragraph are derived from Hellenthal et al., "Genetic Atlas of Human Admixture History."

19. Statement by Alan Goodman in the transcript from episode 1 of the PBS series *Race: The Power of an Illusion*, https://www.pbs.org/race/000_About/002_04 -about-01-01.htm (accessed April 19, 2014).

20. Ibid. (accessed July 15, 2012).

21. United States Census Bureau, "Race," http://www.census.gov/topics/population/race.html (accessed June 25, 2014).

22. P. N. Ossorio, "Myth and Mystification: The Science of Race and IQ," in *Race and the Genetic Revolution: Science, Myth, and Culture*, eds. S. Krimsky and K. Sloan (New York: Columbia University Press, 2011).

23. Ibid. The portions in quotation marks are quoted by Ossorio from N. J. Smelser, W. J. Wilson, and F. Mitchell, eds., introduction to *America Becoming: Racial Trends and Their Consequences, v. 1* (Washington, DC: National Academy Press, 2001), p. 4.

Epilogue

1. R. Bowden et al., "Genomic Tools for Evolution and Conservation in the Chimpanzee: *Pan troglodytes ellioti* Is a Genetically Distinct Population," *PLoS Genetics* 8, no. 3 (2012): e1002504.

2. D. N. Livingstone, "The Preadamite Theory and the Marriage of Science and Religion," *Transactions of the American Philosophical Society*, New Series, 82, no. 3 (1992): 1–78.

3. N. Wade, *A Troublesome Inheritance: Genes, Race and Human History* (New York: Penguin, 2014), pp. 121–22.

4. N. A. Rosenberg et al., "Clines, Clusters, and the Effect of Study Design on the Inference of Human Population Structure," *PLoS Genetics* 1, no. 6 (2005): e70, http://www.plosgenetics.org/article/info%3Adoi%2F10.1371%2Fjournal.pgen .0010070 (accessed September 14, 2014).

5. G. Coop et al., letter to the editor, *New York Times*, August 10, 2014, http:// www.nytimes.com/2014/08/10/books/review/letters-a-troublesome-inheritance.html ?_r=1 (accessed September 14, 2014).

6. T. Dobzhansky, "Nothing in Biology Makes Sense except in the Light of Evolution," *American Biology Teacher* 35 (1973): 125–29.

7. Gallup. "Evolution, Creationism, Intelligent Design," http://www.gallup .com/poll/21814/evolution-creationism-intelligent-design.aspx (accessed June 22, 2014).

BIBLIOGRAPHY

Allison, A. C. "Protection Afforded by Sickle-Cell Trait against Subtertian Malarial Infection." *British Medical Journal* 1, no. 4857 (1954): 290–94.

American Kennel Club. "Breed Matters." https://www.akc.org/breeds (accessed May 3, 2014).

American Society of Hematology. "Statement on Screening for Sickle Cell Trait and Athletic Participation." http://www.hematology.org/Advocacy/Statements/2650.aspx (accessed June 25, 2014).

Australian Government, Department of Immigration and Border Protection. "Fact Sheet 8. Abolition of the 'White Australia' Policy." http://www.immi.gov.au/media/fact-sheets/08abolition.htm (accessed April 19, 2014).

Behar, D. M. et al. "A 'Copernican' Reassessment of the Human Mitochondrial DNA Tree from Its Root." *American Journal of Human Genetics* 90, no. 4 (2012): 675–84.

Beleza, S. et al. "The Timing of Pigmentation Lightening in Europeans." *Molecular Biology and Evolution* 30, no. 1 (2013): 24–35.

Benyamin, B. et al. "Childhood Intelligence Is Heritable, Highly Polygenic and Associated with *FNBP1L*." *Molecular Psychiatry* 19, no. 2 (2014): 253–58.

Billings, P. R. et al. "Discrimination as a Consequence of Genetic Testing." *American Journal of Human Genetics* 50, no. 3 (1992): 476–82.

Bowden, R. et al. "Genomic Tools for Evolution and Conservation in the Chimpanzee: *Pan troglodytes ellioti* Is a Genetically Distinct Population." *PLoS Genetics* 8, no. 3 (2012).

British Broadcasting Corporation (BBC). "Episode 3, Fatal Impacts." *Racism: A History*. http://topdocumentaryfilms.com/racism-history (accessed June 25, 2014).

Brody, H., and L. M. Hunt. "BiDil: Assessing a Race-Based Pharmaceutical." *Annals of Family Medicine* 4, no. 6 (2006): 556–60.

Centers for Disease Control and Prevention. "Adult Cigarette Smoking in the United States: Current Estimates." http://www.cdc.gov/tobacco/data_statistics/fact_sheets/adult_data/cig_smoking/#state (accessed January 12, 2013).

Centers for Disease Control and Prevention. "U. S. Cancer Statistics: An Interactive Atlas." http://apps.nccd.cdc.gov/DCPC_INCA/DCPC_INCA.aspx (accessed January 12, 2013).

Cequeira, C. C. et al. "Predicting *Homo* Pigmentation Phenotype through Genomic Data:

From Neanderthal to James Watson." *American Journal of Human Biology* 24, no. 5 (2012): 705–709.

Chabris, C. F. "IQ Since 'The Bell Curve.'" *Commentary* 106 (1998): 33–40. http://www .wjh.harvard.edu/~cfc/Chabris1998a.html (accessed February 5, 2013).

Chabris, C. F. et al. "Most Reported Genetic Associations with General Intelligence Are Probably False Positives." *Psychological Science* 23, no. 11 (2011): 1314–23.

Chambers, I. et al. "Functional Expression Cloning of *Nanog,* a Pluripotency Sustaining Factor in Embryonic Stem Cells." *Cell* 113, no. 5 (2003): 643–55.

Chin, G., and E. Culotta. "The Science of Inequality: What the Numbers Tell Us." *Science* 344, no. 6186 (2014): 818–21.

Clemens, J. D. et al. "ABO Blood Groups and Cholera: New Observations on Specificity of Risk and Modification of Vaccine Efficacy." *Journal of Infectious Disease* 159, no. 4 (1989): 770–73.

Clements, N. "The Tasmanian Black War: A Tragic Case Lest We Remember?" *The Conversation.* http://theconversation.com/tasmanias-black-war-a-tragic-case-of-lest -we-remember -25663 (accessed June 22, 2014).

Coalition for Genetic Fairness "The History of GINA." http://www.geneticalliance.org/ ginaresource.history (accessed January 27, 2013).

"Companion website for 'A Genetic Atlas of Human Admixture History.'" http:// admixturemap.paintmychromosomes.com (accessed February 23, 2014).

Coop, G. et al. Letter to the editor. *New York Times*, August 10, 2014. http://www.nytimes. com/2014/08/10/books/review/letters-a-troublesome-inheritance.html?_r=1 (accessed September 14, 2014).

Cruciani, F. et al. "A Revised Root for the Human Y Chromosomal Phylogenetic Tree: The Origin of Patrilineal Diversity in Africa." *American Journal of Human Genetics* 88, no. 6 (2011): 814–18.

Currat, M. et al. "Molecular Analysis of the Beta-Globin Gene Cluster in the Niokholo Mandenka Population Reveals a Recent Origin of the Beta-S Senegal Mutation." *American Journal of Human Genetics* 70, no. 1 (2002): 207–23.

D'Errico, P. "Jeffrey Amherst and Smallpox Blankets." http://people.umass.edu/derrico/ amherst/lord_jeff.html (accessed April 27, 2014).

Daley, T. C. et al. "IQ on the Rise: The Flynn Effect in Rural Kenyan Children." *Psychological Science* 14, no. 3 (2003): 215–19.

Darwin, C. R. *On the Origin of Species by Means of Natural Selection, or the Preservation of Favoured Races in the Struggle for Life*. London: John Murray, 1859.

___. *On the Origin of Species by Means of Natural Selection, or the Preservation of Favoured Races in the Struggle for Life*. 4th ed. London: John Murray, 1866.

____. *On the Origin of Species by Means of Natural Selection, or the Preservation of Favoured Races in the Struggle for Life*. 5th ed. London: John Murray, 1869.

Davies, G. et al. "Genome-Wide Association Studies Establish That Human Intelligence Is Highly Heritable and Polygenic." *Molecular Psychiatry* 6, no. 10 (2011): 996–1005.

Derenko, M. V. et al. "The Presence of Mitochondrial Haplogroup X in Altaians from South Siberia." *American Journal of Human Genetics* 69, no. 1 (2001): 237–41.

Dobzhansky, T. "Nothing in Biology Makes Sense except in the Light of Evolution." *American Biology Teacher* 35 (1973): 125–29.

Dulik, M. C. et al. "Mitochondrial DNA and Y Chromosome Variation Provides Evidence for a Recent Common Ancestry between Native Americans and Indigenous Altaians." *American Journal of Human Genetics* 90, no. 2 (2012): 229–46.

Edwards, A. W. F. "Human Genetic Diversity: Lewontin's Fallacy." *BioEssays* 25, no. 8 (2003): 798–801.

Ehlers, C. L. "Variations in *ADH* and *ALDH* in Southwest California Indians." *Alcohol Research & Health* 30, no. 1 (2007): 14–17.

Ehlers, C. L. et al. "Linkage Analyses of Stimulant Dependence, Craving and Heavy Use in American Indians." *American Journal of Medical Genetics Part B Neuropsychiatric Genetics* 156B, no. 7 (2011): 772–80.

Elias, S. A. "Late Pleistocene Climates of Beringia, Based on Analysis of Fossil Beetles." *Quaternary Research* 53, no. 2 (2000): 229–35.

Fairbanks, D. J. *Evolving: The Human Effect and Why It Matters*. Amherst, NY: Prometheus Books, 2012.

Fairbanks, D. J. et al. "*NANOGP8*: Evolution of a Human-Specific Retro-Oncogene." *G3: Genes Genomes Genetics* 2, no. 11 (2012): 1447–57.

Fairbanks, D. J., and P. J. Maughan. "Evolution of the *NANOG* Pseudogene Family in the Human and Chimpanzee Genomes." *BMC Evolutionary Biology* 6 (2006): 12.

Faruque, A. S. G. et al. "The Relationship between ABO Blood Groups and Susceptibility to Diarrhea due to *Vibrio cholerae* 0139." *Clinical Infectious Disease* 18, no. 5 (1994): 827–28.

Flynn, J. R. *What Is Intelligence? Beyond the Flynn Effect*. Cambridge: Cambridge University Press, 2007.

Fry, A. E. et al. "Common Variation in the ABO Glycosyltransferase Is Associated with Susceptibility to Severe *Plasmodium falciparum* Malaria." *Human Molecular Genetics* 17, no. 4 (2008): 567–76.

Fu, W. et al. "Analysis of 6,515 Exomes Reveals the Recent Origin of Most Human Protein-Coding Variants." *Nature* 493, no. 7431 (2013): 216–20.

Gallup. "Evolution, Creationism, Intelligent Design." http://www.gallup.com/poll/21814/evolution-creationism-intelligent-design.aspx (accessed June 22, 2014).

Giardina, E. et al. "Haplotypes in *SLC24A5* Gene as Ancestry Informative Markers in Different Populations." *Current Genomics* 9, no. 2 (2008): 110–14.

Glass, R. I. et al. "Predisposition for Cholera of Individuals with O Blood Group: Possible Evolutionary Significance." *American Journal of Epidemiology* 121, no. 6 (1985): 791–96.

Goldenberg, D. M. *The Curse of Ham: Race and Slavery in Early Judaism, Christianity, and Islam*. Princeton: Princeton University Press, 2003.

Goldsmith, J. C. et al. "Framing the Research Agenda for Sickle Cell Trait: Building on the Current Understanding of Clinical Events and Their Potential Implications." *American Journal of Hematology* 87, no. 3 (2012): 340–46.

Gottfredson, L. S. "Mainstream Science on Intelligence: An Editorial with 52 Signatories, History, and Bibliography." *Intelligence* 24, no. 1 (1997): 13–23.

___. "What If the Hereditarian Hypothesis Is True?" *Psychology, Public Policy, and Law* 11, no. 2 (2005): 311–19.

Gould, S. J. *The Mismeasure of Man*. New York: W. W. Norton, 1981.

___. *The Mismeasure of Man*. Rev. ed. New York: W. W. Norton, 1996.

Graham, G. S., and S. H. McCarty. "Sickle Cell (Meniscocytic) Anemia." *Southern Medical Journal* 23 (1930): 598–606.

Green, R.E. et al. "A Draft Sequence of the Neandertal Genome." *Science* 328, no. 7929 (2010): 710–22.

Haitina, T. et al. "High Diversity in Functional Properties of Melanocortin 1 Receptor (*MC1R*) in Divergent Primate Species Is More Strongly Associated with Phylogeny than Coat Color." *Molecular Biology and Evolution* 24, no. 9 (2007): 2001–2008.

Hammer, M. F. et al. "Population Structure of Y Chromosome SNP Haplogroups in the United States and Forensic Implications for Constructing Y Chromosome STR Databases." *Forensic Science International* 164, no. 1 (2006): 45-55.

Head, T. "Interracial Marriage Laws: A Short Timeline History." http://civilliberty.about.com/od/raceequalopportunity/tp/Interracial-Marriage-Laws-History-Timeline.htm (accessed May 14, 2013).

Healy, E. et al. "Functional Variation of *MC1R* Alleles from Red-Haired Individuals." *Human Molecular Genetics* 10, no. 21 (2001): 2397–402.

Hedges, C., and J. Sacco. *Days of Destruction, Days of Revolt*. New York: Nation Books, 2012.

Hellenthal, G. et al. "A Genetic Atlas of Human Admixture History." *Science* 343, no. 6172 (2014): 747–51.

Henn, B. M. et al. "Genomic Ancestry of North Africans Supports Back-to-Africa Migrations." *PLoS Genetics* 8, no. 1 (2012): e1002397.

Herrnstein, R. J., and C. Murray. *The Bell Curve: Intelligence and Class Structure in American Life.* New York: Free Press, 1996.

Hodges, L. "The Bell Curve Is Sending Shock Waves through America." *Times Higher Education*, November 14, 1994. http://www.timeshighereducation.co.uk/story.asp?story Code=154396&sectioncode=26 (accessed February 1, 2013).

Ingman, M. et al. "Mitochondrial Genome Variation and the Origin of Modern Humans." *Nature* 408, no. 6828 (2000): 708–13.

Jablonski, N. G., and G. Chaplin, "Human Skin Pigmentation as an Adaptation to UV Radiation." *Proceedings of the National Academy of Sciences, USA* 107, suppl. 2 (2010): 8962–68.

James, A. "Making Sense of Race and Racial Classification." In *White Logic, White Methods: Racism and Methodology*, edited by T. Zuberi et al., 31–45. Lanham, MD: Rowman and Littlefield, 2008.

Jarvis, J. P. et al. "Patterns of Ancestry, Signatures of Natural Selection, and Genetic Association with Stature in Western African Pygmies." *PLoS Genetics* 8, no. 4 (2012): e1002641.

Jensen, A. R. "How Much Can We Boost IQ and Scholastic Achievement?" *Harvard Educational Review* 39, no. 2 (1969): 165–96.

Jewish Virtual Library. "The Lebensborn Program (1935–1945)." http://www.jewishvirtuallibrary.org/jsource/Holocaust/Lebensborn.html (accessed July 16, 2012).

Johnson, C. N., and B. W. Brook. "Reconstructing the Dynamics of Ancient Human Populations from Radiocarbon Dates: 10,000 Years of Population Growth in Australia." *Proceedings of the Royal Society, Series B* 278 (2011): 3748–54.

Jolie, A. "My Medical Choice." *New York Times*, May 14, 2013. http://www.nytimes.com/2013/05/14/opinion/my-medical-choice.html?_r=0 (accessed January 5, 2014).

Jorde, L. B., and S. P. Wooding. "Genetic Variation, Classification, and 'Race.'" *Nature Genetics* 36 (2004): S28–S33.

Kaplan, N. L., P. O. Lewis, and B. S. Weir. "Age of the *DF508* Cystic Fibrosis Mutation." *Nature Genetics* 8 (1994): 216.

Keinan, A., and A. G. Clark. "Recent Explosive Human Population Growth Has Resulted in an Excess of Rare Genetic Variants." *Science* 336, no. 6082 (2012): 740–43.

King Jr., M. L. "I Have a Dream." *Historic Documents*. http://www.ushistory.org/documents/i-have-a-dream.htm (accessed June 29, 2014).

Kulozik, A. E. et al. "Geographical Survey of Beta-S-Globin Gene Haplotypes: Evidence for an Independent Asian Origin of the Sickle-Cell Mutation." *American Journal of Human Genetics* 39, no. 2 (1986): 239–44.

Lalueza-Fox, C. et al. "A Melanocortin 1 Receptor Allele Suggests Varying Pigmentation among Neanderthals." *Science* 318, no. 5855 (2007): 1453–55.

Lederer, S. E. *Flesh and Blood: Organ Transplantation and Blood Transfusion in 20th Century America*. Oxford: Oxford University Press, 2008.

Lewontin, R. C. "The Apportionment of Human Diversity." *Evolutionary Biology* 6 (1972): 381–98.

____. "Race and Intelligence." *Bulletin of the Atomic Scientists* 26 (1970): 2–8.

Li, H. et al. "Ethnic Related Selection for an ADH Class I Variant within East Asia." *PLoS One* 3, no. 4 (2008): e1881.

Liu, J. et al. "Haplotype-Based Study of the Association of Alcohol Metabolizing Genes with Alcohol Dependence in Four Independent Populations." *Alcoholism: Clinical and Experimental Research* 35, no. 2 (2011): 304–16.

Livingstone, D. N. "The Preadamite Theory and the Marriage of Science and Religion." *Transactions of the American Philosophical Society*, New Series, 82, no. 3 (1992).

Lu, C. "What Causes 'Asian Glow'?" *Yale Scientific*, April 3, 2011. http://www.yalescientific .org/2011/04/what-causes-"asian-glow" (accessed January 5, 2013).

Makova, K., and H. L. Norton. "Worldwide Polymorphism at the *MC1R* Locus and Normal Pigmentation Variation in Humans." *Peptides* 26, no. 10 (2005): 1901–908.

Malyarchuk, B. et al. "The Peopling of Europe from the Mitochondrial Haplogroup U5 Perspective." *PLoS ONE* 5, no. 4 (2010): e10285.

Mateu, E. et al. "Can a Place of Origin of the Main Cystic Fibrosis Mutations Be Identified?" *American Journal of Human Genetics* 70, no. 1 (2002): 257–64.

Matsuo, Y., R. Yokoyama, and S. Yokoyama. "The Genes for Human Alcohol Dehydrogenases Beta-1 and Beta-2 Differ by Only One Nucleotide." *European Journal of Biochemistry* 183, no. 2 (1989): 317–20.

Meisenberg, G. et al. "The Flynn Effect in the Caribbean: Generational Change in Test Performance in Dominica." *Mankind Quarterly* 46 (2005): 29–70.

Mendez, F. L. et al. "Increased Resolution of Y Chromosome Haplogroup T Defines Relationships among Populations of the Near East, Europe, and Africa." *Human Biology* 83, no. 1 (2011): 39–53.

Morral, N. et al. "The Origin of the Major Cystic Fibrosis Mutation (*DF508*) in European Populations." *Nature Genetics* 7, no. 2 (1994): 169–75.

Nakajima, T. et al. "Natural Selection and Population History in the Human Angiotensinogen Gene (*AGT*): 736 Complete *AGT* Sequences in Chromosomes from around the World." *American Journal of Human Genetics* 74, no. 5 (2004): 898–916.

National Humanities Center Resource Toolbox. "On Slaveholders' Sexual Abuse of Slaves: Selections from 19th & 20th Century Slave Narratives." *The Making of African*

American Identity, Vol. I, 1500–1865. http://nationalhumanitiescenter.org/pds/maai/
enslavement/text6/masterslavesexualabuse.pdf (accessed January 15, 2013).

National Public Radio. "Thomas Jefferson Descendants Work to Heal Family's Past."
http://www.npr.org/templates/story/story.php?storyId=131243217 (accessed November 11, 2012).

Neisser, U. et al. "Intelligence: Knowns and Unknowns." *American Psychologist* 51 (1996):
77–101.

Ng, P. C. et al. "Individual Genomes Instead of Race for Personalized Medicine." *Clinical Pharmacology and Therapeutics* 84 (2008): 306–309.

Nisbett, R. E. "Heredity, Environment, and Race Differences in IQ: A Commentary on
Rushton and Jensen." *Psychology, Public Policy, and Law* 11, no. 2 (2005): 302–10.

Nisbett, R. E. et al. "Intelligence: New Findings and Theoretical Developments." *American Psychologist* 67, no. 2 (2012): 129.

Norton, H. L. et al. "Genetic Evidence for the Convergent Evolution of Light Skin in
Europeans and East Asians." *Molecular Biology and Evolution* 24, no. 3 (2007): 710–22.

Omenn, G. S. "Evolution and Public Health." *Proceedings of the National Academy of Sciences, USA* 107, suppl. 1 (2012): 1702–709.

Ossorio, P. N. "Myth and Mystification: The Science of Race and IQ." In *Race and the Genetic Revolution: Science, Myth, and Culture*, edited by S. Krimsky and K. Sloan. New
York: Columbia University Press, 2011.

Pauling, L. et al. "Sickle Cell Anemia: A Molecular Disease." *Science* 110 (1949): 543–48.

Perego, U. A. et al. "Distinctive Paleo-Indian Migration Routes from Beringia Marked by
Two Rare mtDNA Haplogroups." *Current Biology* 19, no. 1 (2009): 1–8.

Plomin, R. "Child Development and Molecular Genetics: 14 Years Later." *Child Development* 84, no. 1 (2013): 104–20.

Plomin, R., and M. Rutter. "Child Development, Molecular Genetics, and What to Do
with Genes Once They Are Found." *Child Development* 69, no. 4 (1998): 1223–42.

Pośpiech, E. et al. "The Common Occurrence of Epistasis in the Determination of Human
Pigmentation and Its Impact on DNA-Based Pigmentation Phenotype Prediction."
Forensic Science International: Genetics 11 (2014): 64–72.

Public Broadcasting System (PBS). "Mapping Jefferson's Y Chromosome." *Frontline.*
http://www.pbs.org/wgbh/pages/frontline/shows/jefferson/etc/genemap.html (accessed November 11, 2012).

Punnett, R. C. *Mendelism.* New York: Macmillan, 1905.

Raghavan, M. et al. "Upper Palaeolithic Siberian Genome Reveals Dual Ancestry of Native
Americans." *Nature* 505 (2014): 87–91.

Race, Ethnicity, and Genetics Working Group. "The Use of Racial, Ethnic, and Ancestral

Categories in Human Genetics Research." *American Journal of Human Genetics* 77, no. 4 (2005): 519–32.

Raper, A. B. "Sickle Cell Disease in Africa and America: A Comparison." *Journal of Tropical Medicine and Hygiene* 53 (1950): 49–53.

Rasmussen, M. et al. "An Aboriginal Australian Genome Reveals Separate Human Dispersals into Asia." *Science* 334, no. 6052 (2011): 94–98.

Reed, A. G. *The Hemingses of Monticello: An American Family*. New York: W. W. Norton, 2009.

Rich, S. M. et al. "The Origin of Malignant Malaria." *Proceedings of the National Academy of Sciences, USA* 106, no. 35 (2009): 14902–907.

Rosenberg, N. A., et al. "Clines, Clusters, and the Effect of Study Design on the Inference of Human Population Structure." *PLoS Genetics* 1, no. 6 (2005): e70, http://www.plosgenetics.org/article/info%3Adoi%2F10.1371%2Fjournal.pgen.0010070 (accessed September 14, 2014).

Rowe, J. A. et al. "Blood Group O Protects against Severe *Plasmodium falciparum* Malaria through the Mechanism of Reduced Rosetting." *Proceedings of the National Academy of Sciences, USA* 104, no. 44 (2007): 17471–76.

Rusert, B. M., and C. D. M. Royal. "Grassroots Marketing in a Global Era: More Lessons from BiDil." *Journal of Law and Medical Ethics* 39, no. 1 (2011): 79–90.

Rushton, J. P., and A. R. Jensen. "Thirty Years of Research on Race Differences in Cognitive Ability." *Psychology, Public Policy, and Law* 11, no. 2 (2005): 235–94.

Savittand, T. L., and M. F. Goldberg. "Herrick's 1910 Case Report of Sickle Cell Anemia: The Rest of the Story." *Journal of the American Medical Association* 261, no. 2 (1989): 266–71.

Schwartz, M., and D. Vissing. "Paternal Inheritance of Mitochondrial DNA." *New England Journal of Medicine* 347, no. 8 (2002): 609–12.

Scott, R. B. "Health Care Priorities and Sickle Cell Anemia." *Journal of the American Medical Association* 214, no. 4 (1970): 731–34.

———. "Reflections on the Current Status of the National Sickle Cell Disease Program in the United States." *Journal of the National Medical Association* 71, no. 7 (1979): 679–81.

Ségurel, L. et al. "The ABO Blood Group Is a Trans-Species Polymorphism in Primates." *Proceedings of the National Academy of Sciences, USA* 109, no. 45 (2012): 18493–98.

Sharma, S. et al. "Vitamin D Deficiency and Disease Risk among Aboriginal Arctic Populations." *Nutritional Review* 69, no. 8 (2011): 468–78.

Sickle Cell Disease Association of America. "Sickle Cell Trait and Athletics." http://www.sicklecelldisease.org/index.cfm?page=sickle-cell-trait-athletics (accessed December 30, 2012).

Smelser, N. J., W. J. Wilson, and F. Mitchell, eds. Introduction to *America Becoming: Racial Trends and Their Consequences, v. 1.* Washington, DC: National Academy Press, 2001.

Smith, R. et al. "Melanocortin 1 Receptor Variants in an Irish Population." *Journal of Investigative Dermatology* 111, no. 1 (1998): 119–22.

Soares, P. et al. "Correcting for Purifying Selection: An Improved Human Mitochondrial Molecular Clock." *American Journal of Human Genetics* 84, no. 6 (2009): 740–59.

South Africa Parliament. *Report of the Joint Committee on the Prohibition of Mixed Marriages Act and Section 16 of the Immorality Act.* Cape Town, South Africa: Government Printer, 1985.

South African History Online. "The Battle of Blood River." http://www.sahistory.org.za/dated-event/battle-blood-river (accessed April 28, 2014).

Spearman, C. S. "'General Intelligence,' Objectively Determined and Measured." *American Journal of Psychology* 15, no. 2 (1904): 201–92.

Stern, A. M. *Eugenic Nation: Faults and Frontiers of Better Breeding in Modern America.* Oakland, CA: University of California Press, 2005.

Sternberg, R. J. "Intelligence." *Dialogues in Clinical Neuroscience* 14, no. 1 (2012): 19–27.

___. "There Are No Public-Policy Implications: A Reply to Rushton and Jensen." *Psychology, Public Policy, and Law* 11, no. 2 (2005): 295–301.

Sternberg, R. J., E. L. Grigorenko, and K. K. Kidd. "Intelligence, Race, and Genetics." *American Psychologist* 60, no. 2 (2005): 176.

Stringer, C. B. et al. "ESR Dates for the Hominid Burial Site of Es Skhul in Israel." *Nature* 338 (1989): 756–58.

Sulem, P. et al. "Genetic Determinants of Hair, Eye and Skin Pigmentation in Europeans." *Nature Genetics* 39 (2007): 1443–52.

Summerhayes, G. R. et al. "Human Adaptation and Plant Use in Highland New Guinea 49,000 to 44,000 Years Ago." *Science* 330, no. 6000 (2010): 78–81.

Tapper, M. *In the Blood: Sickle Cell Anemia and the Politics of Race.* Philadelphia, PA: University of Pennsylvania Press, 1999.

Thomas Jefferson Memorial Foundation. *Report of the Research Committee on Thomas Jefferson and Sally Hemings.* http://www.monticello.org/sites/default/files/inline-pdfs/jefferson-hemings_report.pdf (accessed November 11, 2012).

Tishkoff, S. A. et al. "Convergent Adaptation of Human Lactase Persistence in Africa and Europe." *Nature Genetics* 39 (2007): 31–40.

Trzaskowski, M. et al. "DNA Evidence for Strong Genetic Stability and Increasing Heritability of Intelligence from Age 7 to 12." *Molecular Psychiatry* 19, no. 3 (2014): 380–84.

United States Census Bureau. "Race." http://www.census.gov/topics/population/race.html (accessed June 25, 2014).

United States Holocaust Memorial Museum. "Holocaust Encyclopedia." http://www
.ushmm .org/wlc/en/article.php?ModuleId=10005143 (accessed July 16, 2012).

Van Oven, N., and M. Kayser. "Updated Comprehensive Phylogenetic Tree of Global
Human Mitochondrial DNA Variation." *Human Mutation* 30, no. 2 (2009): E386–94.

"Voyages." *Trans Atlantic Slave Trade Database*. http://www.slavevoyages.org/tast/index.
faces (accessed December 24, 2013).

Wade, N., *A Troublesome Inheritance: Genes, Race and Human History*, New York: Penguin,
2014.

Wagner, C. R., F. R. Greer, and the Section on Breastfeeding and Committee on Nutrition.
"Prevention of Rickets and Vitamin D Deficiency in Infants, Children, and Adolescents." *Pediatrics* 122, no. 5 (2008): 1142–52.

Warren, E. "*Loving v. Virginia*: Opinion of the Court." (No. 395) 206 Va. 924, 147 S.E.2d 78,
reversed.   http://www.law.cornell.edu/supct/html/historics/USSC_CR_0388_0001_
ZO .html (accessed July 15, 2012).

Watson, M. S. et al., eds. "Newborn Screening: Toward a Uniform Screening Panel and
System." *Genetics in Medicine* 8, suppl. 1 (2006): 1S–252S.

Williams, L. K. et al. "Differing Effects of Metformin on Glycemic Control by Race-Ethnicity." *Journal of Clinical Endocrinology and Metabolism* (early release, in press, 2014).
http://press .endocrine.org/doi/abs/10.1210/jc.2014-1539 (accessed June 25, 2014).

Witt, S. H. "Pressure Points in Growing up Indian." *Perspectives* 12, no. 1 (1980): 24–31.

Wojcicki, A. "23andMe Provides an Update Regarding FDA's Review." *23andMe Blog*,
December 5, 2013. http://blog.23andme.com/news/23andme-provides-an-update-
regarding-fdas-review (accessed January 5, 2013).

Xing, J. et al. "Fine-Scaled Human Genetic Structure Revealed by SNP Microarrays."
Genome Research 19 (2009): 815–25.

Zeng, F. Y. et al. "Sequence of the −530 Region of the Beta-Globin Gene of Sickle Cell
Anemia Patients with the Arabian Haplotype." *Human Mutation* 3, no. 2 (1994)
163–65.

Zhang, J. et al. "Genomewide Distribution of High-Frequency, Completely Mismatching
SNP Haplotype Pairs Observed to Be Common across Human Populations." *American Journal of Human Genetics* 73, no 5 (2003): 1073–81.

Zoutendyk, A., A. C. Kopec, and A. E. Mourant. "The Blood Groups of the Hottentots."
American Journal of Physical Anthropology 13, no. 4 (1955): 691–97.

INDEX